Introduction to Proteomics

INTRODUCTION TO PROTEOMICS

Sudhir Awasthi

ARCLER
PRESS

www.arclerpress.com

Introduction to Proteomics

Sudhir Awasthi

Arcler Press
224 Shoreacres Road
Burlington, ON L7L 2H2
Canada
www.arclerpress.com
Email: orders@arclereducation.com

ISBN: 978-1-77469-509-8 (Hardcover)

ABOUT THE AUTHOR

Prof. Sudhir K. Awasthi did his M.Sc in Life Sciences securing first position in CSJM University, Kanpur. Thereafter he completed his PhD in the same discipline from CSJM University, Kanpur and joined the faculty of Life Sciences as Asstt. Professor in 1991. He became Professor in 2008 & presently working as Pro- Vice Chancellor & Dean , Faculty Life Sciences in this university . He is also holding the post of Director , Deendayal Shodh Kendra of the University campus from past one year he is also occupying the post of Co-ordinator, Recruitment & Assessment Cell (RAC) responsible for the recruitment & assessment of university faculty members. Prof. Awasthi has more than 30 years of teaching & research experience. His area of research has been Immuno-toxicology & Environmental Biochemistry. He has been widely recognized as one of those few researchers who did excellent work in the field of Mycobacterial Dormancy & the role of Heat Shock Proteins / Chaperons in its granuloma state. He has published more than 40 research papers & completed two research projects funded by DST & UPCST respectively. He is Principle Investigator of Centre of Excellence Project for Department of Life Sciences funded by UP Council of Higher Education. He is reckoned as prolific orator & popular article writer in different domains of knowledge related to higher education. He has been successful in advocating the importance of traditional Indian Knowledge System which is being re- explored in present context. He has participated in several conferences as Lead Speaker & Chaired the session in numerous seminars & symposia.

TABLE OF CONTENTS

LIST OF FIGURES

LIST OF TABLES

PREFACE

With the previous decade, a technological revolution culminated in the sequencing of the human genome and the beginning of the postgenomic era. In the early 1990s, Marc Williams and his colleagues created the phrases proteomics and proteome, which have subsequently been used by the research community at large. The proteome is the collection of all proteins encoded by the genome. Historically, biochemical investigations of protein function have concentrated on analyzing single molecular species. The fast identification of novel gene products through large-scale genomic and proteomic programs has forced the development of alternate methods for assessing protein activity. In recent years, the goal has been to create high-throughput methods to permit systematic protein analysis in biological samples, to map functional connections between proteins on a global scale, and to position them in a biological context.

This book introduces the emerging topic of proteomics. It aims to describe how proteins and proteomes may be researched and assessed. In spite of the increased interest in proteomics, the scientific community as a whole is only slowly gaining knowledge of proteomics methods and technology. This book tackles the need to introduce biologists to new tools and methods and is intended for both biology students and experienced biologists in practice. This book should provide anyone who has attended a graduate-level biochemistry course with a reasonable knowledge of what proteomics is and how it is conducted. Much of this should be recognizable to the seasoned biologist, but refocused to assist proteome research.

Long-sought benchmarks in genome sequencing, analytical apparatus, computing power, and user-friendly software tools have fundamentally altered the profession of biology. After years of researching the separate components of biological systems, it is now possible to examine the systems as a whole in molecular detail. We must thus efficiently deploy new technology, cope with mounds of data, and, most importantly, adapt our thinking to comprehend complex systems as opposed to their component parts.

The book is divided into eight chapters. The first chapter introduces the readers to the fundamentals of Proteomics. Chapter 2 deals with the discussion of proteomics with respect to genome and genomics. Chapter 3 thoroughly discusses the methodology for the separation and identification of proteins and their interactions.

Chapter 4 introduces the readers to information on chromatography and its application in protein studies. Chapter 5 focuses on proteomics of protein modifications. Chapter 6 illustrates the phenomena of proteomics in protein-protein interactions. Chapter 7 describes the modern applications of proteomics in various fields. Finally, chapter 8 focuses on the recent developments in the field of proteomics and protein studies.

The book provides an excellent overview of the many facets of the proteome. The text succeeds in presenting the information in a manner that familiarizes the inexperienced reader with the important concepts and tools of proteomic. *Introduction to Proteomics* is equally beneficial for students of physics and other multidisciplinary fields.

-Author

CHAPTER 1

FUNDAMENTALS OF PROTEOMICS

CONTENTS

1.1 INTRODUCTION

The word "proteome" is derived from the terms genome and protein. It reflects the whole assortment of proteins encoded by an organism's DNA. Thus, proteomics is expressed as the total protein composition of an organism or a cell. Proteomics is the study of the function, structure, and relationships of an organism's complete protein composition. Proteins regulate the phenotypic of a cell by defining its structure and, more importantly, by performing all cellular activities. Diseases are mostly caused by defective proteins, which serve as important markers for the identification of a specific disease. Proteins are the basic targets of the majority of pharmaceuticals and serve as the basic foundation for the creation of novel medications. Thus, the research of proteomics is essential to comprehend their involvement in the etiology and management of illnesses, and also in the growth of children and other species (Alsagaby, 2019).

Although RNA is used by some viruses, DNA has served as the medium for the transmission of proteins in the vast majority of organisms. DNA, apart from RNA viruses, has been transcribed into RNA, which has been then always translated into a protein. RNA viruses are an exception to this rule (Arruda et al., 2011; Xia, 2018). When it comes to RNA viruses, RNA undergoes direct translation into proteins. At one point in time, it was thought that a single gene would generate a single enzyme, which would then regulate a phenotypic trait. The discovery that eukaryotic genes are split into two parts, which results in RNA splicing, the possibility of RNA editing, and the phenomenon of RNA silencing, has caused substantial alterations in this perspective over the last few decades. The fragmented nature of genes, RNA splicing, RNA silencing, and RNA editing are all topics that are going to be discussed later on in this section (Bousquet et al., 2011).

In eukaryotes, noncoding regions of nucleotides known as introns disrupt the coding patterns of a gene known as exons (Zhang et al.,2008). Upon elimination of introns, exons have been spliced either consistently (known as cis splicing) or intermittently (known as alternative splicing) inside a gene, or between exons of other genes (trans-splicing). Protein abundance in eukaryotic organisms is a result of the many types of exon splicing and post-translational modifications of proteins. There have been around 23,000 genes and over 500,000 proteins in humans (Barlow et al., 2008; Bonetta& Valentino, 2020).

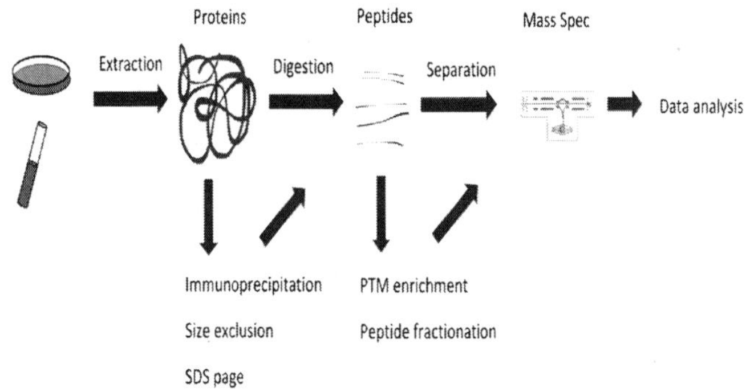

Figure 1.1. Fundamental Proteomics Workflow.

Source: https://health.usf.edu/medicine/corefacilities/proteomics/introduction

The finding that each gene encodes for a single enzyme may appear to conflict with recent findings on the existence of suppressor genes and the division of genes. It was not until the establishment of fundamental dogma in biology (Crick, 1958, 1970, Watson 1965, Mattick 2003, Lewin 2004) and the deciphering of the genetic code (Leder and Nirenberg 1964, Khorana 1968) that it became apparent how suppressor genes carry out their functions. Therefore, the method of action of suppressor genes does not in any way contradict the essential assumptions suggested in Beadle and Tatum's one-gene-one-enzyme model. It is understandable, given the fundamental dogma, that certain genes or parts of DNA can code for a variety of proteins, or that the coding region of the protein in DNA is scattered throughout a large area of DNA and is broken up by noncoding sequences. It is already common knowledge that the concept of "one-gene–one-enzyme" relates exclusively to genes that code for a single polypeptide and does not apply to genes that have a split nature and are capable of producing several proteins. Therefore, the assumption that one gene encodes for only one enzyme is restricted to the nature of genes, just as the Mendelian principles of inheritance only apply to genes that are placed in the nucleus of the cell and do not apply to genes that are located in any other portion of the cell (Watson et al., 2005; Huberts& van der Klei, 2010). Therefore, Mendelian inheritance relates to the arrangement of the genes, whereas the assumption that "one-gene–one-enzyme" is limited to the type of gene.

What Tatum and Beadle proposed is a rule, not an axiom, and certain circumstances are just exceptions to their deep rule. Nature, it appears, follows the British dictum that "exceptions establish the rule." Exceptions abound in the history of science. The basic dogma in molecular biology stated by Francis Crick, the co-discoverer of the DNA structure, is the most obvious case of such an exception. Crick (1958, 1970) hypothesized that sequential data in DNA is transmitted to RNA, which is subsequently transferred to protein, and that the direction of this data transmission is fixed. Moreover, it was later discovered that RNA is reverse transcribed into DNA and that messenger RNA (mRNA) is sometimes edited before translation into protein by adding or removing cytidine or uridine, implying that data in a DNA fragment has not been translated directly into protein as implied by central dogma. This theory proposes that DNA produces RNA, which then produces protein. The Nobel Prize was awarded to Howard Temin and David Baltimore in 1975 for proving the reverse transfer of information from RNA to DNA. The enzymes are another obvious instance of such an exception. The discovery that enzymes are proteins was made by Cornell University's James Sumner. Soon after, Sydney Altman of Yale University and Thomas Cech of the University of Colorado separately demonstrated that some enzymes are comprised of RNA rather than proteins. Sumner received the Nobel Prize in 1946, while Altman and Cech received it in 1989 for their work in chemistry. As a result, it appears that biology, like any other discipline of study, is full of exceptions to the laws (Nabieva et al., 2005; Lee et al., 2007).

Berzelius (1838)1, a Swedish chemist, classified several naturally present polymers proteins. Sumner stated the fact that enzymes are proteins (1946). Proteins are formed up of patterns of amino acids, according to Sanger (1958). Linus Pauling showed in the 1940s that a substrate and an enzyme (or an antibody and an antigen) must have a perfect complementary fit in their structures to interact with each other, similar to a hand in a glove. Aside from Sumner (1946), both Sanger (1958) and Pauling (1954) were awarded Nobel Prizes for their works in chemistry. Although most proteins serve as enzymes, some, like fibronectin and actin, serve as structural elements of cells. Muscle, bones, and cartilage are all made up of proteins. Proteins are also essential for muscle cell motility. Proteins can act in a variety of ways, such as receptors for other molecules, as immunoglobulins or antigens, as allergens, or as components in the transport of other substances, such as oxygen or sex hormones. All of these roles can be played by proteins. Several proteins are hormones that control critical metabolic functions in humans and other creatures, like insulin and human growth hormone (HGH). Proteins' 3-dimensional structure and chemical changes are critical to understanding their functioning in many capacities (Thomas et al., 2003; Jiang et al., 2007).

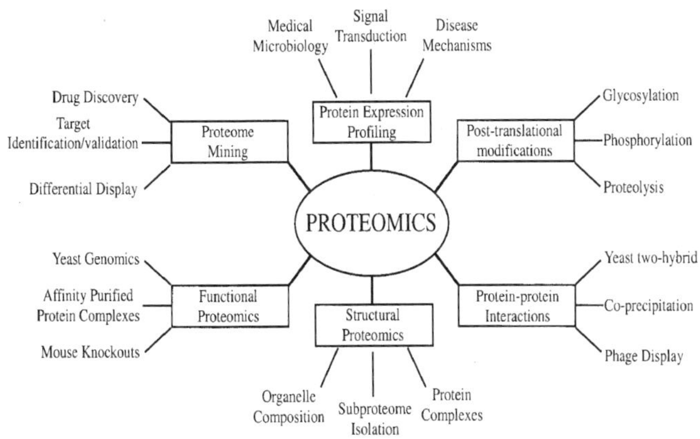

Figure 1.2. Proteomics kinds and its biological applications.

Source: https://pharmaceuticalintelligence.com/2014/11/07/introduction-to-proteomics/

Gorrod (1909) was the first to identify various human problems as inborn metabolic defects, implying a genetic foundation for such illnesses. However, it was the brilliance of Tatum and Beadle (1941) that resulted in the discovery that a gene encodes a protein. They demonstrated that a mutant's synthesis of a component in a metabolic pathway was hampered by working with Neurospora. They demonstrated that when a gene directing an enzyme that catalyzes a biochemical step in a metabolic pathway was disabled, the mutant acquired dietary needs for such substance. These mutants might not grow on a minimum media, but only when a specific chemical was introduced to the minimal medium did they grow. A mutant having impaired arginine production, for instance, might not grow on a minimal medium and might only grow when arginine was given to the minimal medium. The biochemical pathways were also mapped using this method (Schlessinger et al., 2007; Hinman& Lou, 2008).

Tatum and Beadle (1941) referred to this theory as the "one gene-one enzyme hypothesis." This concept has been reworked in several different ways. Nevertheless, the one-gene-one-enzyme concept has maintained the cornerstone of the biological discipline despite several notable outliers. This strategy was essential in bringing chemistry and genetics closer together, which was a prerequisite for the development of molecular biology. Producing a mutant and demonstrating which protein lacks a function or which function is hindered in a specific protein is the standard method for assigning a function to a protein. Because of this theory, it was feasible to do study and analysis on the genetics of viruses, plants, microbes, and animals. The development of knockout mutations and mutagenesis *in vitro* have both been built on top of this basis. This theory has shown to be a vital tool for studying any fundamental genetic mechanism, such as recombination, repairing, and replication of DNA, as well as for determining the role that a protein plays in any metabolic pathway. In the end, Beadle and Tatum's concept was responsible for breakthroughs made in the fields of agriculture, medicine, and pharmaceutical sciences. The one-gene-one-enzyme hypothesis is significant in the research and treatment of human diseases, as well as the development of gene therapy. This theory also served as the basis for the theory (Alberts et al., 2002; Ajit Tamadaddi &Sahi, 2016).

According to the one-gene–one-enzyme concept, a mutation should have changed the protein. Due to a lack of available technology at the time, Tatum and Beadle were unable to illustrate the faulty character of the mutant protein. Yanofsky(1952, 2005a,b) and Lein and Mitchell (1948, Mitchell, et al. 1948) first proved this at the biochemical level using mutants of Neurospora lacking the enzyme essential for tryptophan production. This hypothesis was demonstrated previously at the molecular level by Ingram (1957) in the example of sickle cell anemia patients' hemoglobin. Ingram demonstrated that the 6th amino acid "glutamic acid," which is present in the hemoglobin molecule of a healthy individual, is replaced by valine in a patient's hemoglobin molecule. This single shift from glutamic acid to valine is the cause of sickle cell blood diseases. Subsequently, it was discovered that other additional mutants lacked proteins entirely or possessed proteins with changed amino acid sequences.

There is a correlation between the ordered location of nucleotides in a gene and the organized position of amino acids in the protein that is made by that gene, which is another thing that is represented by the one-gene-one-enzyme theory. Both Yanofsky et al. (1964) and Sarabhai et al. (1964) independently demonstrated that there is a colinear relationship between the structure of a gene and that of a protein. This colinearity will be explored in further detail in the following sections (Lill, 2009).

1.2 PROTEINS THAT PERFORM PARTICULAR FUNCTIONS

The proteome is the total number of proteins encoded by an organism's DNA. Proteomics is the study of characterizing an organism's proteome classification and characteristics.

Marc Wilkins coined the phrase "proteome" in 1994. (Wilkins 1996). O'Farrell (1975) and Klose (1976) both attempted to characterize the total proteins of an organism (1975). By performing gel electrophoresis of proteins in 2 planes at right angles to one another, they invented two-dimensional gel electrophoresis (O'Farrell 1975, Klose 1975). On the gel, our approach sorted a complicated mixture of greater than 1100 *Escherichia coli* proteins into discrete bands of individual elements. Subsequently, the application of mass spectrometry in combination with genomics for the large-scale identification and isolation of proteins transformed the field of proteomics.

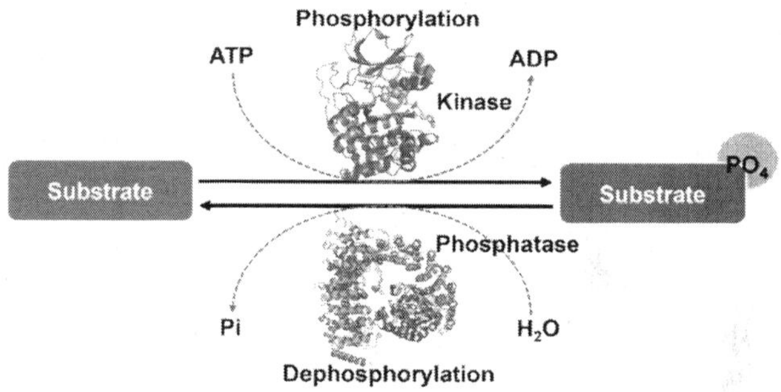

Figure 1.3. Phosphorylation may make a protein inactive or active.

Source: Seok S-H. Structural Insights into Protein Regulation by Phosphorylation and Substrate Recognition of Protein Kinases/Phosphatases. Life. 2021; 11(9):957. https://doi.org/10.3390/life11090957

An organism's genome is stable in the sense that it is identical in all types of cells at all times. The proteome of an organism, on either hand, is dynamic since it varies from one type of cell to the next and changes even within the identical type of cell at various times of development or activity. A change in the proteome reflects differences in gene activity that are reliant on the type of cell to produce the protein required for a certain function. Blood cells, for instance, mostly exhibit the hemoglobin gene, which creates the hemoglobin protein essential for oxygen transport, while pancreatic cells primarily exhibit the insulin gene, which generates the insulin peptide needed for glucose entry into cells (Neuwald et al., 1999; Song & Singh, 2009).

Since every protein controls a particular function, variable gene expression is necessary for the creation of diverse proteins. Table 1.1 lists the functions of numerous proteins. Furthermore, a cell's protein profile might alter based on the numerous types of protein modifications, which may include phosphorylation, acetylation, glycosylation, or interaction with lipid or carbohydrate molecules. Protein enhancements happen as a result of post-translational processes and change protein function. The mitosis activator protein (MAP) kinase protein, for instance, controls mitosis and is phosphorylated to produce MAP Kinase Kinase (MAPKK), MAP Kinase (MAPK), and MAP Kinase Kinasekinase (MAPKKK). The significance of protein enhancement in cellular activity regulation has been examined (Pellegrini, 2003; Marschalek, 2011).

1.3 PREGENOMIC PROTEOMICS

Protein functioning as enzymes in regulating cellular activity was well understood even before their structure had been discovered. The enunciation of the one-gene enzyme notion provided the conceptual insight into interpreting the structure of a protein as a linear arrangement of amino acids. Several technological advancements enabled this conceptual breakthrough (Axe, 2004; Rittmann et al., 2008).

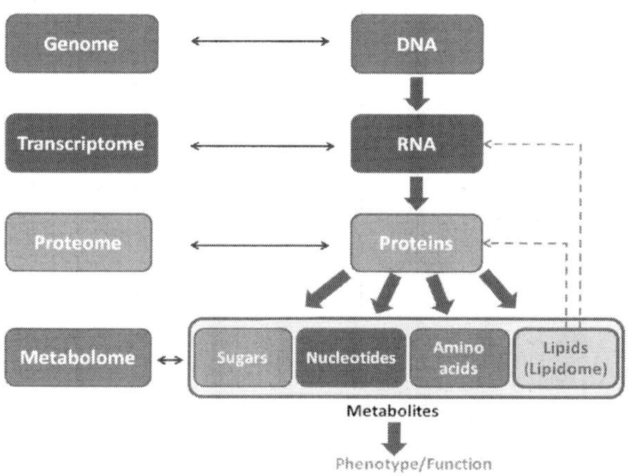

Figure 1.4. Metabolomics Schema.

Source: https://en.wikipedia.org/wiki/File:Metabolomics_schema.png

The technological advancements included the creation of equipment for analyzing the amino acid content and determining the amino acid sequence of a protein. Using such machines, the structure of proteins was determined by protein for many years. Subsequently, the development of the two-dimensional gel technique and mass spectrometry made it possible to simultaneously resolve the structures of many proteins (Shishkin et al., 2004; Paape &Aebischer, 2011).

Table 1.1. The function of different proteins

Protein	Function
Enzymes (greater than ninety percent of proteins) Catalyze biochemical reactions in the cell.	Catalyst
Hemoglobin (carrier of oxygen) Albumin (carrier of hormones)	Transport
Cartilage/bone proteins	Structure
Actin, fibrinoactin	Cellular skeleton
Insulin, a growth hormone	Hormone
Immunoglobulins	Antibody
Bacterial and viral proteins	Antigens and allergens
Myosin	Mobility/muscle movement
Receptor for cholesterol	Receptors
Transduction proteins, junction proteins	Cell communication/signaling

With the advent of bioinformatics and genomics, it became possible to understand the shape of multiple proteins at the same time using mass spectrometry. The nucleotide pattern of DNA/genes in the chromosomes of many creatures was decoded using genomics approaches. Bioinformatics approaches entailed studying the majority of an organism's nucleotide sequence using computers and a variety of software tools. Bioinformatics is also utilized to deduce a protein's amino acid sequence from the nucleotide sequence of a DNA molecule (D'Hooghe et al., 2004; Chen et al., 2016).

1.4 GENETICS OF PROTEINS

The one-gene–one-enzyme concept necessitated a genetic model for understanding the function and structure of proteins. This notion indicated that the function and structure of proteins might be comprehended by comparing proteins derived from wild-type animals and mutant species. Understanding the involvement of a protein in either metabolic or developmental path is now standard practice. Following this rule, the hemoglobin molecules of healthy persons and sickle cell sufferers had been compared. The 6th amino acid in normal persons' hemoglobin is distinct from that of sickle cell sufferers. Normal persons have glutamic acid at this location, while the sickle cell sufferer have valine (Ingram 1956, 1957). Therefore, a single amino acid substitution entirely transformed the function and structure of hemoglobin (Vandahl et al., 2004; Jurkiewicz et al., 2009).

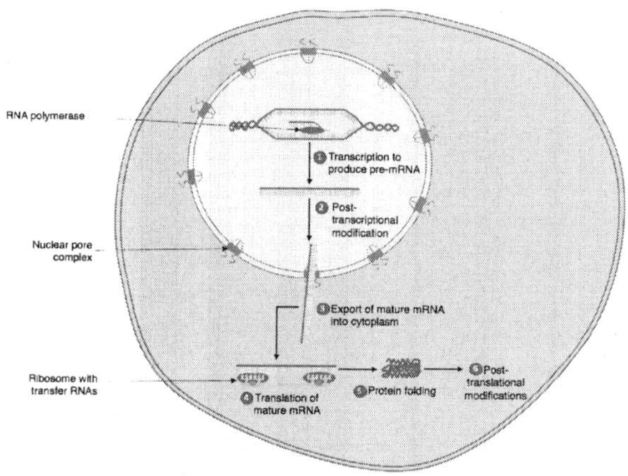

Figure 1.5. The Process of Protein Synthesis.

Source: By Kep17 - Own work, CC BY-SA 4.0, https://commons.wikimedia.org/w/index.php?curid=89042835 One-Gene-One-Enzyme Theory

According to the idea that was presented by Tatum and Beadle (1941), the structure of an enzyme or protein is thought to be controlled by a single gene. This idea is based on the fact that a single gene will only ever encode for a single protein. This concept proved to be useful for understanding the biology of any metabolic pathway as well as the function of proteins that were responsible for triggering the biochemical reaction at each stage of that process.

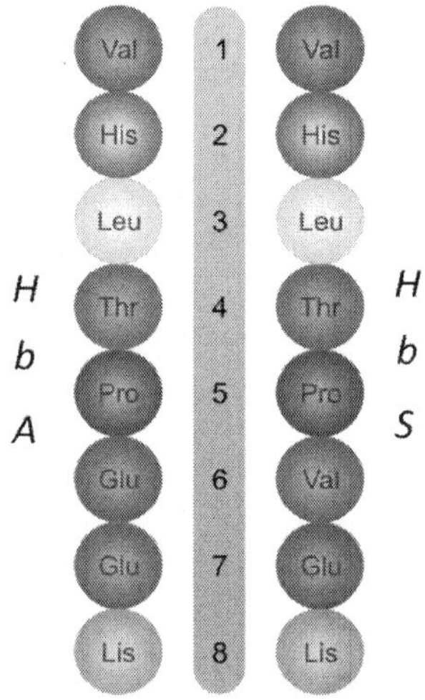

Figure 1.6. The N-terminal amino acid pattern of the beta chain of sickle and normal cell hemoglobin was compared.

Source: https://commons.wikimedia.org/wiki/File:Hbs.svg

In the beginning, it was discovered that if an organism is unable to develop without the assistance of a supplement, such as a nucleotide, a specific amino acid, or a vitamin, then that organism is lacking in the protein that catalyzes the biochemical reaction that is responsible for the synthesis of that substance, which is now a nutritional necessity for its growth (Chemale et al., 2006; Li et al., 2016).

This resulted in the development of a technique for identifying mutants with a particular dietary demand and the sequence of biological processes in a metabolic path. This investigation of nutritional mutants indicated the existence of a distinct category of mutants. It had been discovered that a subset of mutants needs the amino acids ornithine, arginine, or citrulline for development. The 3rd set of mutants might only develop in the existence of arginine, but a second group needed either arginine or citrulline for development (Colozza et al., 2007). The nutritional needs of this last group of mutants were not addressed by adding citrulline or ornithine to the development media, either alone or in combination. The dietary needs of such 3 mutant groups revealed a metabolic route for the organism's manufacture of arginine. This metabolic route consisted of biochemical processes including the sequential production of ornithine from a precursor molecule, citrulline from ornithine, and arginine from citrulline. Consequently, the following metabolic passageway was founded:

Precursor →Ornithine→Citrulline→Arginine.

It is clear, based on the order of the biochemical activities that take place in this route, that the first set of mutants has a problem with the step that involves the transformation of the precursor into ornithine. As a result, this population of mutants is capable of obtaining the nutrients necessary for development from either citrulline, ornithine, or arginine. The 2nd set of mutants is deficient in the process that involves the transformation of ornithine into citrulline. As a consequence of this, the requirement for development in this group of mutants can potentially be satisfied by the introduction of citrulline or arginine, but not ornithine. The third group of mutants has a problem in the very last phase of the biochemical process, which is the moment in which citrulline is converted into arginine. As a consequence of this, the only way for an organism to mature is if arginine is supplied in the form of a supplement. As a consequence of this, the idea of "one gene-one enzyme" eventually developed into a useful tool for determining the sequential order of biological events that take place along a certain passageway. This theory also claimed that if the enzyme that catalyzes the transition of substance A into substance B is incorrect for whatever reason, then the molecules of substance A would gather within the organism. A potential threat to the health of mutants can emerge if an excessive amount of this substance is allowed to build up in their bodies.

The buildup of phenylalanine in phenyl-ketoneurics and the buildup of homogentisic acid in newborns with alcaptonuria demonstrate this. These metabolic bottlenecks arise in the phenylalanine–tyrosine metabolic route as a consequence of certain enzyme abnormalities. Gorrod referred to these genetic flaws as "inherited metabolic disorders" (1909). An excess of phenylalanine can harm the brain's growth in the initial phases of growth, perhaps leading to mental impairment. In the US and other industrialized nations, now it is essential to test newborns after delivery for phenyl-ketoneuria by looking for an elevated level of phenylalanine in the blood. To control phenylalanine levels, phenyl-ketoneuric newborns are put on a particular protein-deficient diet. Such children are reintroduced to a regular diet after their brain growth is completed. Therefore, a phenylketoneuric woman should limit her phenylalanine consumption during pregnancy for the baby's brain to develop properly (Choi et al., 1999; Hu, 2000).

By comparing the biophysical characteristics of the wild-type and mutant enzymes participating in a certain metabolic passageway, this idea was later utilized to demonstrate the identification of a specific protein and its participation in a biochemical step.

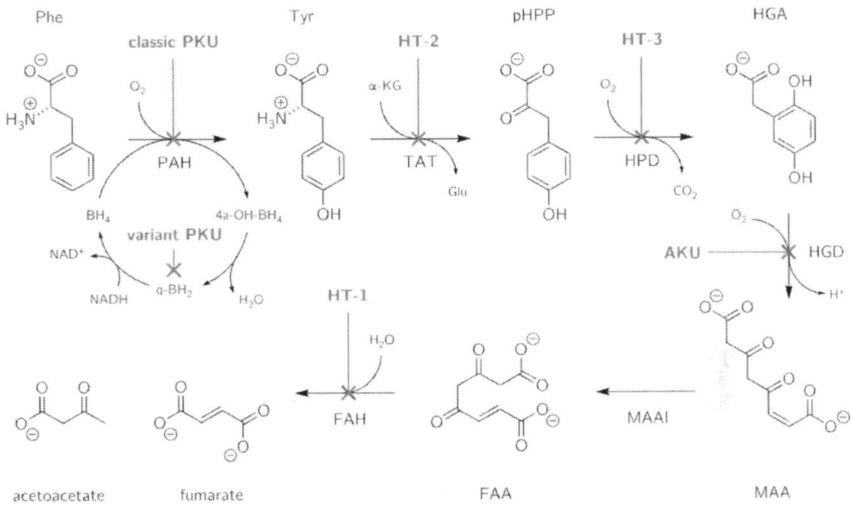

Figure 1.7. Outcomes of a phenylalanine-tyrosine metabolic bottleneck in phenylketonuric newborns, a faulty phenylalanine hydroxylase may cause a build-up of phenylalanine, that may damage brain cells and cause mental disabilities. Alkaptonuria is caused by a metabolic obstruction induced by a faulty enzyme.

Source: By Bradford Morris - Own work, CC BY-SA 4.0, https://commons.wikimedia.org/w/index.php?curid=78901270

It was immediately determined that a mutant either did not produce a particular protein, produced only a portion of that protein, or produced an imperfect protein that included a particular amino acid in a particular location. The presence of distinct subtypes of mutant proteins is directly proportional to the nature of the transformations that take place as a result of alterations to the genetic code. Examples of these alterations include changing the position of one nucleotide in the genetic code relative to another, as well as removing or adding a nucleotide in the order of the gene's DNA. A single nucleotide can be changed within the genetic code, which can then lead to a mutation in the protein that is either missense, nonsense, or silent. In the process of converting an existing amino acid codon into a stop codon, a nonsense mutation can take place. If there is a nonsense mutation at the beginning of a gene that codes for the protein, the resulting protein will either be a short peptide or nothing at all. Any mutation in the gene that is considered to be nonsense would produce truncated proteins of varying lengths. A protein's biochemical properties can be altered through a mutation known as missense, which involves the replacement of single amino acid with another. This can result in the protein being active or partially active. In addition, because of degeneracy in the genetic code or the fact that a replacement amino acid cannot affect the overall function and structure of the protein, the replacement of a single nucleotide in the genetic code with another nucleotide cannot produce any change in the protein that is produced. These mutations are referred to be silent or neutral mutations. A shift in the reading of the triplet genetic code occurs when a nucleotide is deleted or inserted in the genetic code. From the point of addition or removal of the nucleotide, a frameshift mutation causes changes like all amino acids. If it happens at the start or in the center of the gene, then it causes alterations in a high number of the amino acids that are produced, which renders the protein entirely inactive. If it happens at the end of the gene, then it has no effect. Conversely, if the addition or removal of a nucleotide arises toward the end of the gene, it is feasible that the resultant amino acid changes will still leave the function of the protein unaffected. This is because the end of the gene has fewer nucleotides than the beginning. It has been discovered that all such different sorts of mutations can happen in the genome of an organism (Lee et al., 1995; Colozza et al., 2007).

According to the one-gene-one-enzyme approach, a mutant will either be deficient in a certain protein or lack that protein entirely. This had been demonstrated initially with a mutant of Neurospora that had been dependent on tryptophan, and afterwards with analogous mutants of *E. coli.* At the moment, dozens of mutants are investigated, which demonstrates the one-to-one existence of a relationship between genes and proteins. Mutants usually have either no protein or a faulty protein that is devoid of enzymatic activity. Therefore, the one-gene-one-enzyme theory offers not only an important role in the understanding of the gene in the procedure of encoding a protein, but it also supplied a tool for dissecting the biochemistry of every simple to complicated procedures in the living system. This was accomplished by generating mutants and afterwards comparing the biochemical changes that have taken place in the mutant. There is not a single system that is not within the scope of this potent instrument (Kniemeyer, 2011; Muthuirulan, 2016).

1.4.1 Colinearity of Gene and Protein

The one-gene-one-enzyme idea of Tatum and Beadle (1941) laid the groundwork for colinearity in the protein structures and DNA/gene by proposing that the gene reflects a nucleotide pattern and the protein reflects the amino acids sequence. By transfecting bacteria and bacterial viruses, Avery et al. (1944) and Hershey and Chase (1952) demonstrated that genes are composed of DNA molecules. The correspondence between the genetic maps of particular mutants and nucleotide blocks demonstrated that the gene is a pattern of nucleotides. The investigation of missense mutants of *E. coli* (Yanofsky et al., 1964) or nonsense mutants of a bacterial virus revealed this colinearity between the amino acid pattern of proteins and the DNA sequence of genes (Sarabhai et al. 1964).

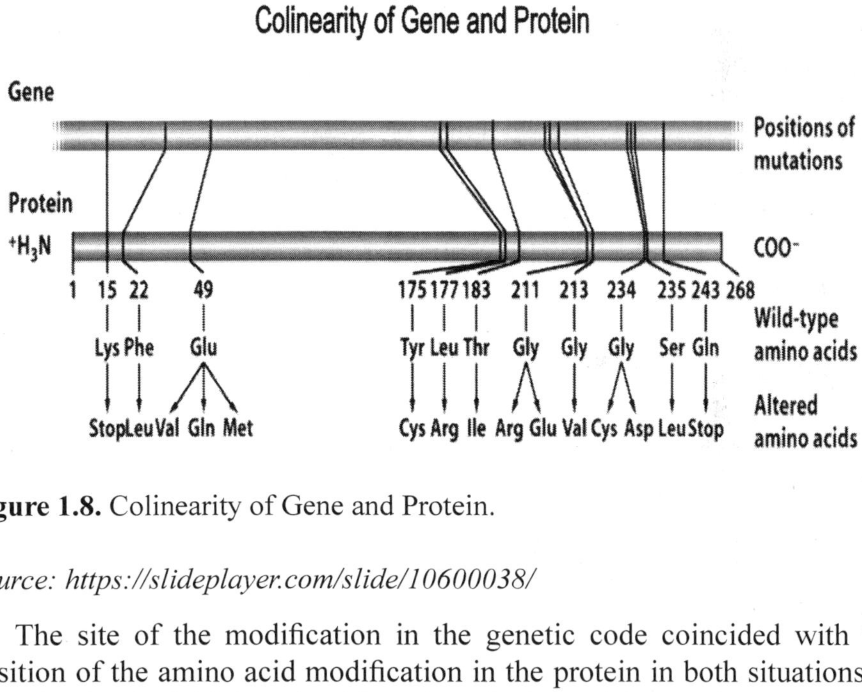

Figure 1.8. Colinearity of Gene and Protein.

Source: https://slideplayer.com/slide/10600038/

The site of the modification in the genetic code coincided with the position of the amino acid modification in the protein in both situations. A mutation in the initial nucleotide order of a bacterial gene for protein was discovered by Yan of Sky et al. A mutation in the gene's middle correlated to a mutation in the protein's middle amino acid position. A modification in the terminus of a gene correlated to a mutation in the location of the protein toward the end. An enzyme that synthesizes tryptophan. Sarabhai et al. (1964) demonstrated that a virus produces shortened viral proteins; the size of the peptides is related to the length of the gene, and sense modification ensued (Florens et al., 2002; Sun et al., 2017).

1.4.2 Protein as a Sequence of Amino Acids

By deciphering the structure of insulin polypeptide as a linear series of various amino acids, Sanger showed beyond a reasonable doubt that a protein is in reality an order of amino acids (1958). As a result, insulin had been the 1st polypeptide or tiny protein to be sequenced, and Sanger did it using his method (1958). Stein and Moore (1972) were the first people to sequence a full-size protein when they did so with ribonuclease A, which was an enzyme (Biron et al., 2007; Yao et al., 2018).

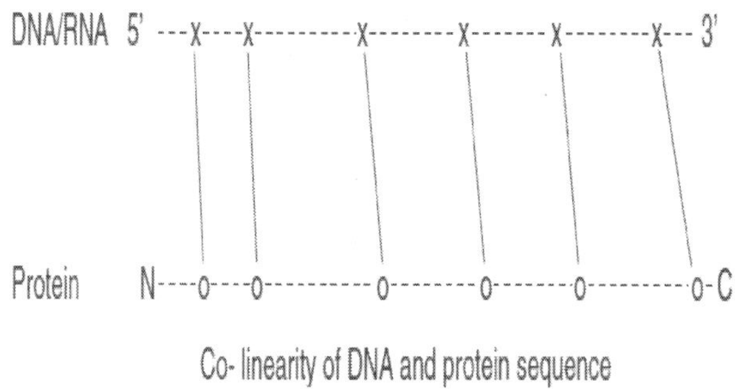

Figure 1.9. Sequence similarity between protein and DNA. The X indicates the mutation location in the gene or DNA as determined by recombinational analysis. The O indicates the location of changed amino acids within the gene-encoded protein. Vertical lines link the locations of mutations in the protein and gene to illustrate their exact relationship.

Source: https://www.nature.com/scitable/topicpage/what-is-a-gene-colinearity-and-transcription-430/

When the technology for cloning a gene as well as its sequence analysis became accessible, although, the direct evidence that a gene is an order of nucleotides had been achieved. Before a 3-dimensional structure is established, proteins typically have four types of structure. Such distinct structures are referred to as primary, secondary, tertiary, and quaternary structures. The main structure of proteins is the linear order of amino acids. The tertiary and secondary structures occur from the interaction between the side groups connected to the amino acids and the folding of polypeptide upon itself. Two or more completely folded polypeptides link with one another to form the quaternary structure of the proteins (Sunagar et al., 2016; Taha et al., 2020).

The one-gene–one-enzyme concept implied that the basic structure of the peptide defines the quaternary, tertiary, and secondary structure, and this had been formed by Anfinsen (1973) through a study of mutant ribonuclease and chemical change, and the renaturation and denaturation kinetics of such an enzyme (Anfinsen1973).

1.5 ONE GENE-MANY PROTEIN: CHALLENGE TO PROTEOMICS

The basic dogma of biology implies that genetic data flows from DNA to RNA to protein in the order of DNA-RNA Protein.

The Tatum and Beadle one-gene-one-enzyme theory is expressed as follows in this system: One DNA (Transcript or mRNA) a single protein So the protein-encoding data in bacterial genes is continuous and the transcript is immediately translatable and comparable to mRNA, this technique works well for them. Furthermore, it had been quickly discovered that several genes in eukaryotes had a split gene structure, with noncoding regions (intron) interrupting the protein-encoding sequences (exon). Because several eukaryotic genes are divided, the transcript should be processed to remove noncoding intervening regions (introns) and make all coding portions or exons continuous to produce mRNA that can be translated. Exon splicing may occur in a variety of ways, resulting in distinct types of mRNA from the same transcript (Loscalzo, 2011; Bludau&Aebersold, 2020).

As a result of the split structure of eukaryotic genes, the Tatum and Beadle idea of the gene-enzyme relationship must be amended, as one gene may produce several proteins, and might be described in core dogma language as

One DNA→one transcript→several mRNAs→several proteins

It's worth noting that when it had been discovered that RNA might be reverse transcribed into DNA, the core dogma shifted. The central dogma has been renamed.

DNA↔RNA→Protein, instead of DNA→RNA→Protein

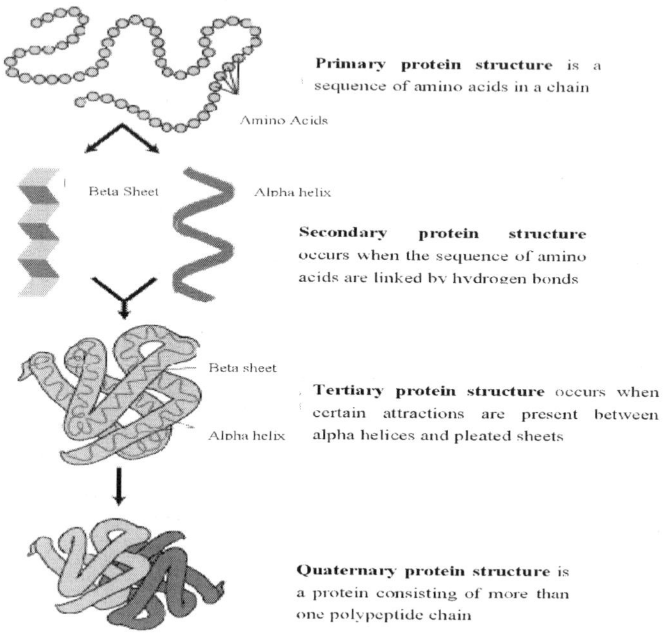

Figure 1.10. Protein structure with various degrees of order. Darryl Leza of NIHGR/NIH granted permission for this reproduction.

Source: https://www.quora.com/What-is-the-difference-between-the-disordered-structure-of-a-protein-and-loop-structure

As a result, the core dogma is no longer an axiom, which also applies to Tatum and Beadle's one-gene-one-enzyme idea. They do embody several important biological principles. Therefore, when new facts about the nature of genes emerge, such rules must be updated to accommodate them (Chaijaroenkul et al., 2014; Makjaroen et al., 2018).

The novel concept that a single gene may encode several proteins has aided in comprehending how a human with just 23,000 genes can encode over 90,000 proteins. Humans were estimated to contain 100,000 or more genes in the pre-genomic era. The human genome project, on the other hand, revealed the existence of approximately 23,000 genes that encode proteins. This apparent contradiction is resolved by the principle that although each gene only produces a single transcript, each transcript can give rise to a large number of mRNAs. These mRNAs are then translated into a variety of proteins. As a consequence of this, the 23,000 human genes may encode all of the body's 90,000 proteins. In higher eukaryotic organisms like primates (including humans) and rodents, more than 50% of the genes encode several proteins (Lander et al. 2001). It has been hypothesized that the Drosophila DSCAM gene is responsible for the encoding of more than 38,000 different proteins. It is estimated that there are approximately 500,000 different proteins present in different human cells at different stages. This increase in the number of proteins found in human cells is caused by post-translational modifications of the 90,000 proteins that are encoded by the 23,000 human genes. Furthermore, it's worth noting that nearly every gene in prokaryotes encodes only one protein (Kuhn & Goebel, 2006; Matheson et al., 2015).

Over ninety percent of genes in lower eukaryotic organisms, like filamentous or yeast fungus, encode for just one protein each. This image is completely different in higher creatures, such as humans, in which more than fifty percent of genes only encode one protein per gene while other genes encode many proteins per gene. It would appear that one gene may code for more than 3 distinct proteins in higher eukaryotic organisms on average (Neidhardt et al., 1984; Shewry et al., 2003).

1.5.1 RNA Splicing

A gene must first undergo transcription into a transcript or pre-mRNA before it can be used by higher organisms. To create mRNA that can be translated the latter goes through a series of further alterations known as "processing." There has been a minimum of three phases involved in the procedure. In the 1st phase, a cap or a new guanosine nucleotide is added at the 5'end. In the 2nd phase, a tail or poly-A nucleotides is added at the 3'end. Both of these phases are optional. The 3rd phase in the process involves removing from the transcript any intervening noncoding sequences that are known as introns. The process of RNA splicing includes the elimination of introns and the merging of exons to ensure that the various coding sequences included within a transcript are continuous in the mRNA that has been produced. A combination of proteins and RNAs that is structured into an organelle known as splicosomes are responsible for carrying out the process of RNA splicing. A splicosome is comparable in size to a ribosome and serves as a platform for the joining of exons and the elimination of introns. This process takes place on the surface of the splicosome. Several consensus sequences, like GA at the 5'end and GU at the 3'end of the intron, are used to identify the two ends of an intron. An intron is a segment of DNA that is located between 2 exons. Throughout the procedure of RNA splicing, an intron will loop out and then be eliminated as a lariat structure. This structure will have a guanine nucleotide serving as the tail, which will link the adjacent exons together. Certain introns can eliminate themselves by a process known as self-splicing, which does not require a splicosome. Only in eukaryotic organisms can the pre-mRNA transcript undergo the process of RNA splicing. Furthermore, several transfer RNAs (tRNAs) can undergo splicing in both eukaryotes and prokaryotes. The splicing of these tRNAs is performed by specific enzymes and does not require splicosomes in any way (Payne et al., 1984; Garattini et al., 2003).

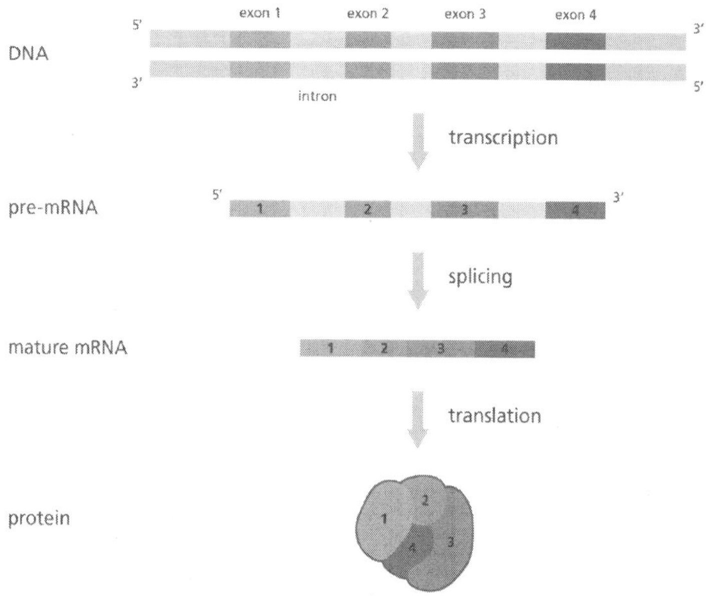

Figure 1.11. A diagram depicting the RNA splicing procedure.

Source: https://www.yourgenome.org/facts/what-is-rna-splicing

Pre-mRNA in eukaryotes can be processed in a range of methods. The multiple exons of a pre-mRNA are 1st brought together consistently by removing introns, resulting in one translatable mRNA. A pre-mRNA with 3 exons and 2 introns, for instance, would create an mRNA with all 3 exons together when the introns are removed; this mRNA would translate into a lengthy protein. Secondly, the many exons of this or comparable pre-mRNAs can go through alternative splicing, resulting in multiple translatable mRNAs. For instance, a pre-mRNA containing 3 exons and 2 introns can perform alternative splicing, resulting in 2 distinct messages: one mRNA having one and two exons combined, as well as the other mRNA having one and three exons together. As a result, during translation, these 2 mRNAs would create separate proteins. Exons from 2 distinct pre-mRNAs can sometimes be spliced together to produce distinct mRNAs. Trans splicing is a type of splicing which includes the exons of separate pre-mRNAs (Payne, 1987; Floros& Hoover, 1998).

Alternate splicing is the primary reason for the creation of several proteins from a single gene. Trans-splicing results in the production of single or more proteins from two genes. These two situations are a significant deviation from Tatum and Beadle's original one-gene-one-enzyme idea (1941). Although, it appears reasonable at the molecular level as proteins and enzymes are formed up of modules transcribed by exons. Nature has created methods to bring such modules together, like alternative trans-splicing and splicing, to generate a functional protein or enzyme (Maestri et al., 2002; Emilsson et al., 2018).

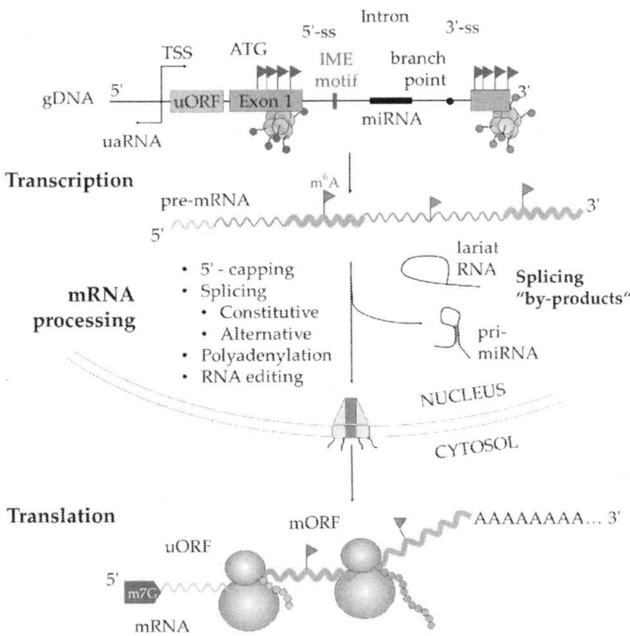

Figure 1.12. Intron deletion from a transcript.

Source: Zalabák D, Ikeda Y. First Come, First Served: Sui Generis Features of the First Intron. Plants. 2020; 9(7):911. https://doi.org/10.3390/plants9070911

1.5.2 RNA Editing

The editing of RNA, in addition to the splicing of RNA, seems to be another factor, which alters the nature of proteins. By the editing of RNA, a single gene can create several functional proteins. As a result, the editing of RNA may have an impact on an organism's proteome. The deletion or insertion of a cytidine or uridine nucleotide from the mRNA produces a change in the character of the codon in the mRNA before it is translated, which is known as RNA editing. The deletion or addition of a nucleotide while RNA editing is aided by an RNA termed as guide RNA (gRNA). Organellar mRNA is frequently edited. RNA can be modified in various ways besides deletion or insertion editing, like the transformation of cytidine to uridine or the transformation of adenosine to inosine by particular deaminases. Conversion editing is the term for such procedures. When adenosine is transformed into inosine, it is interpreted as guanosine by the ribosome. As a result, a CAG codon for glutamine becomes CGG after inosine conversion, and it encodes for arginine rather than glutamine. Apart from tRNA, mRNA, micro RNA (miRNA), and ribosomal (rRNA) can all be edited. The editing of tRNA usually results in the reading of a stop codon into leucine (Chiti et al., 2002; Rasheed et al., 2014).

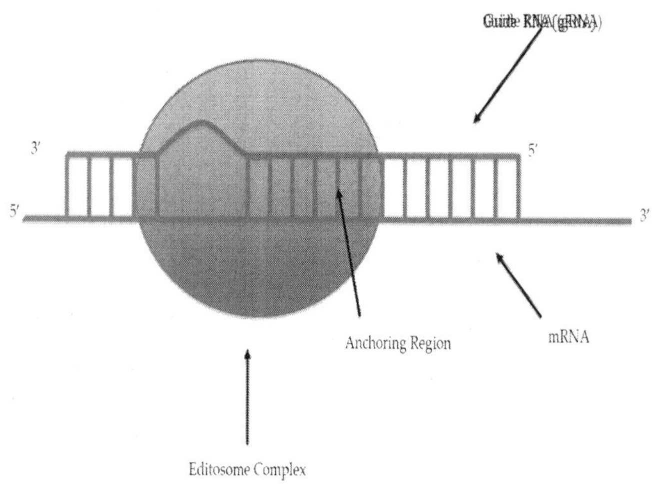

Figure 1.13. RNA editing is depicted graphically.

Source: https://www.sciencedirect.com/topics/neuroscience/rna-editing

Not only does the procedure of the editing of RNA alter the protein's nature, but it also provides a counterexample to something called the central dogma. This is because it shows that data may be transferred directly from DNA to RNA into proteins.

RNA editing demonstrates that proteins are produced from information not found in the DNA sequence in some cases. Lou Gehrig and Human cancer disease, also known as amyotrophic lateral sclerosis (ALS), have been linked to poor RNA editing (Begun et al., 2000).

1.6 RNA SILENCING AND PROTEOMICS

In animals, fungi, and plants, a whole new method for gene regulation has been discovered in current years. This method involves triggering the breakdown of mRNA to silence a particular gene-specific signal. The expression of the resident gene(s), viral-induced gene(s), transgene(s), and transposons is controlled by RNA silencing (Bishop et al., 2000). Whenever a gene for anthocyanin had been inserted to elaborate the pigment or color development in the petunia flower, this had been discovered initially. In these tests both the resident and foreign transgenes for the synthesis of color were inhibited, and the plant rather generated white flowers. Post-transcriptional gene silencing (PTGS) had been coined to describe this phenomenon. Gene repression was later discovered in the fungus Neurospora. The insertion of an orange color gene into Neurospora led to albino or white transformants, according to the findings. Quelling is the process of silencing a gene, like a pigment synthesis gene in Neurospora. The white or albino Neurospora transformants didn't create mRNA particular for the color gene, according to the findings. The quelling of the resident gene had also been revealed to be limited to a portion of the transgenic having up to 130 nucleotides in length and not the complete color gene. Whenever a heterokaryon had been created between the transgenic and wild-type strains of Neurospora, the transgene had been observed to quell or inhibit the expression of resident genes while in another nucleus. Subsequently, quelling-deficient Neurospora mutant strains had been discovered; such mutants had been dubbed "quelling deficient" (qde). In Neurospora, there have been 3 classes of these mutations. RNA-dependent RNA polymerase (RdRP) is important for the formation of double-stranded RNA dsRNA like siRNA or miRNA throughout the silencing of the gene in Neurospora.

The Sting or Piwi class of proteins, which are linked to the translational factor eIF2C, are encoded by the qde-2 gene. The Neurospora qde-3 gene codes for a WRN (Warner's syndrome) protein having DNA or RNase helicase functions that are comparable to RecQ DNA helicase.

Various organisms, like fission yeast, worms (*Caenorhabditis elegans*), and Arabidopsis, have been shown to have counterparts of Neurospora qde-3, qde-2, and qde-1, genes. RdRP family proteins are identified in a variety of plants, such as wheat, tomato, fission yeast, petunia, and *C. elegans*. This protein is capable of synthesizing a complimentary copy of gene-particular mRNA. That copy of RNA becomes a two-stranded RNA when it combines with mRNA. An enzyme known as Dicer, which is related to RNase III ribonuclease, degrades the latter into smaller RNA pieces. The RNA segments attach to an RNA-induced silencing complex (RISC) and break the mRNA specific to a gene, resulting in gene silence or suppression (Aschaffenburg, 1968; Setlow, 1988).

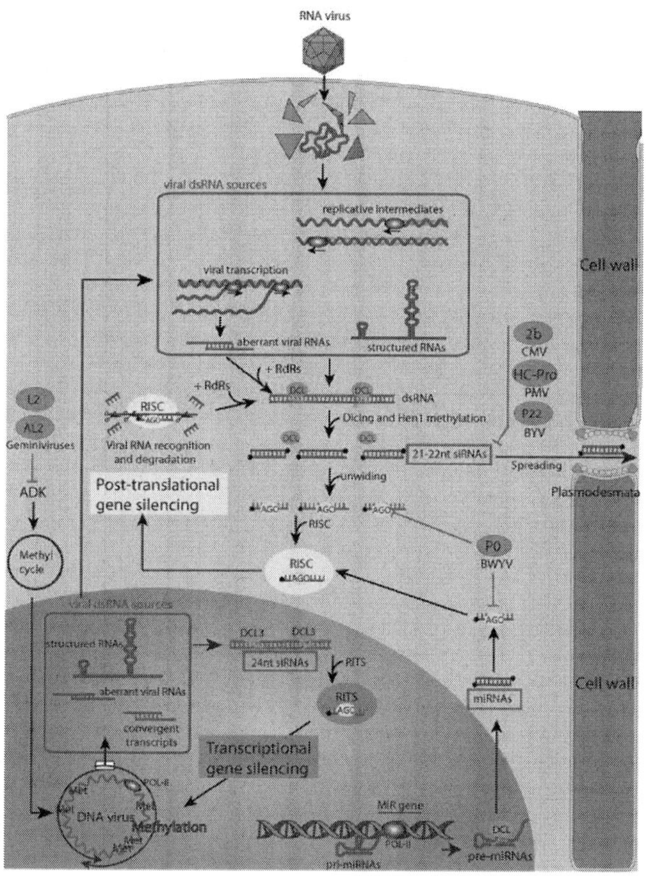

Figure 1.14. RNA Silencing is depicted graphically.

Source: https://viralzone.expasy.org/891?outline=all_by_species

Experiments using Drosophila or worms revealed the importance of dsRNA in silencing, and it has been now detected in cells of Mammalia too. The introduction of little fragments of dsRNA unique to a gene has been proven to cause the disintegration of its mRNA, resulting in the reduction of that gene's expression. As a result, RNA silencing may be utilized to modify gene expression in organisms, and it has the potential to be a powerful tool in the treatment of a variety of human disorders, such as cancer. In 2006, Mello and Fire were awarded the Nobel Prize for discovering the process of RNA interference (Fire et al. 1998). Several infectious organisms, such as trypanosomes, intestinal parasites, and viruses are called to cause havoc in humans due to their antigenic differences. Moreover, it has been now recognized that some human intestinal parasites, like *Giardia lamblia*, utilize RNA interference to sustain antigenic diversity (Prucca et al. 2008). Knowing this mechanism could lead to better control of various human viral diseases (Snipes & Suter, 1995; Ecker et al., 2008).

1.7 MOLECULAR BIOLOGY OF GENES AND PROTEINS

A gene is a stretch of nucleotide order or a DNA segment that, via translation or transcription, encodes a protein. Certain genes, although, produce RNA that is not translated into proteins. Throughout transcription, such genes produce tRNA or rRNA, that aid in the interpretation of transcripts from protein-coding genes. The coding section of DNA in bacterial genes is uninterrupted, and their transcripts have been translated straight into protein without alteration. Therefore, mRNA and transcript are synonymous in prokaryotes (such as RNA, which transmits the instructions for building a protein through the ribosome translation procedure). Astrachan and Volkin (1957) confirmed the presence of mRNA in the cells of bacteria, and Brenner et al. proposed that mRNAs convey data from DNA to ribosomes for interpretation into proteins (1961). H. G. Khorana and Marshall Nirenberg discovered the genetic information as well as the process of data transfer and storage predicted by the Watson-Crick structure of DNA at the same time. The Nobel Prize was awarded to Nirenberg and Khorana in 1968 for its accomplishments. In the early 1960s, the variable size of eukaryotic transcripts had been recognized. The eukaryotic transcripts are known as pre-messenger RNA (premRNA) or heterogeneous nuclear RNA (hnRNA). Consequently, the finding of the separate nature of eukaryotic genes demolished the notion of the diverse nature of eukaryotic transcripts. Midway through the 1970s, it became evident that certain genes in eukaryotes contain separate structures wherein the coding DNA compartment termed exons are separated by the noncoding DNA compartment termed introns. This finding had been derived from the findings of heteroduplex mapping, which involved the hybridization of a gene's DNA using mRNA and the electron microscope imaging of the heteroduplex structure. Whenever the DNA of a gene had been hybridized with the mRNA in these kinds of tests, specific DNA sequences displayed as loops (Sharp, 2005). The existence of such loops suggested the existence of introns, which had been lacking from the mRNA. Based on the outcomes of such hybridization experiments, it had been determined that a transcript experiences splicing processes that result in the removal of introns. Therefore, the exons have been rendered continuous, and the message is only then capable of being translated. In eukaryotes, such findings showed a contrast between the structure of a transcript and its mRNA. Subsequently, the existence of introns and exons in a gene had been established by comparing the chicken ovalbumin gene's DNA order and its mRNA. Upon completion of the genome projects of numerous animals, the existence of introns and exons in a gene may be readily determined by locating conserved nucleotides at the exon-intron junctions (Payne et al., 1982; Ferré et al., 1995).

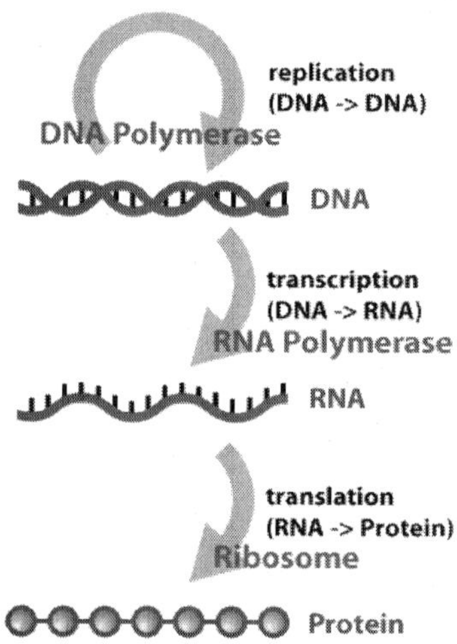

Figure 1.15. DNA commands are converted into messenger RNA. Ribosomes can read the genetic data encoding a strand of mRNA and utilize it to link amino acids together to produce a protein.

Source: By Dhorspool at en.wikipedia, CC BY-SA 3.0, https://commons.wikimedia.org/w/index.php?curid=15183788

At first, it was believed that introns are effectively removed from a transcript through the process of splicing, which then causes the exons to form a continuous sequence in the mRNA. Later on, it was discovered that a single transcript might result in the production of a wide variety of mRNAs. This was made possible by two distinct processes known respectively as alternative and trans-splicing. During the process known as alternative splicing, the exons are pieced back together in a variety of various configurations. For instance, if a gene has three exons, one mRNA could comprise exon 1 and exon 2, while another messenger RNA from a similar gene could possess exon 1 and exon 3. This would cause the two messenger RNAs to generate completely distinct proteins having unique amino acid sequences at the C-terminal ends of the proteins. The functions controlled by such two proteins in an organism's physiology will be completely distinct from one another and involve the regulation of various biochemical reactions. Therefore, based on the total number of exons in a gene, this form of alternative splicing has the potential to generate a wide variety of messenger RNA for unique proteins (Akam et al., 1978; Andersen et al., 2021).

The DSCAM gene in Drosophila is thought to create over 38,000 messenger RNAs encoding various proteins. Because of a variety of factors, such as the existence of an initial stop codon, not all messenger RNAs are translatable. Alternate splicing may be tissue-specific, resulting in proteins having specialized functions. The Bcl-x gene produces a protein that regulates apoptosis, or programmed cell death. Furthermore, this gene uses a distinct splicing method to produce two separate mRNAs. A mini version of messenger RNA generates the protein Bcl-x(s), which regulates cancer and enhances apoptosis, while a bigger version of messenger RNA generates a large protein that increases cancer growth and reduces apoptosis.

Intron retention or exon skipping may occur during alternative splicing. Exon skipping occurs often in higher eukaryotes. While exon skipping, a certain exon is omitted from the splicing process. There have been multiple instances of exon skipping, that are utilized to generate variants of tropomyosin that have been unique to smooth muscle, skeletal muscle, and the cells of the brain. Drosophila utilizes exon skipping to regulate sexual maturation. Drosophila possesses a sex lethal gene known as sxl; if exon 2 is bypassed while splicing, a female-specific sxl protein is created, that connects with all succeeding transcripts of the similar gene, which induces excision of exon 2 from all mRNAs, and results in the generation of female flies. If the male-specific exon 2 is kept during the 1st round of splicing, the male-specific sxl protein is produced, and male flies hatch. The retention of introns leads to the synthesis of messenger RNAs and their encoding proteins of various lengths. The retention of introns is a typical occurrence in plants and lower multicellular animals (Virgin et al., 1991; Caestecker& Meyrick, 2001).

One transcript has always been involved in the alternate splicing method. Trans splicing, in contrast to alternative splicing, comprises the splicing of exons from 2 transcripts generated by similar or different genes. In worms like *C. elegans*, trans-splicing is very common. Trans splicing can happen in human brain cells, according to certain evidence.

A typical human gene that codes for a protein are around 28,000 nucleotides long, has approximately eight exons that are at least 120 nucleotides long and has approximately seven introns that range in size between 100 and 100,000 nucleotides. Introns typically have a length that is many times greater than that of exons. Alternate splicing results in the production of an average of three different mRNAs from each human gene (Baker et al., 1966; Nasrallah et al., 1970).

Splicosomes which are composed of greater than 100 proteins and 5 short nuclear (sn) RNAs (snRNAs) are responsible for facilitating the splicing process. Splicing regulator proteins, also known as SR proteins, are regulatory proteins that engage splicosomes by binding to a specific nucleotide order in an exon known as an exon splicing enhancer (ESE). The exon may have an order known as an exon splicing suppressor (ESS), that stops the splicosome from doing its job (Swanson & Bradfield, 1993; Barisoni et al., 2009).

The humans might get sick from improper splicing. Defective splicing is the outcome of more than fifteen percent of the mutations that are responsible for illnesses in humans. Splicing errors can lead to mutations that modify the splice site, the elements of splicesosomes, or the factors that drive splicing. Mutations can also alter the factors that influence splicing.

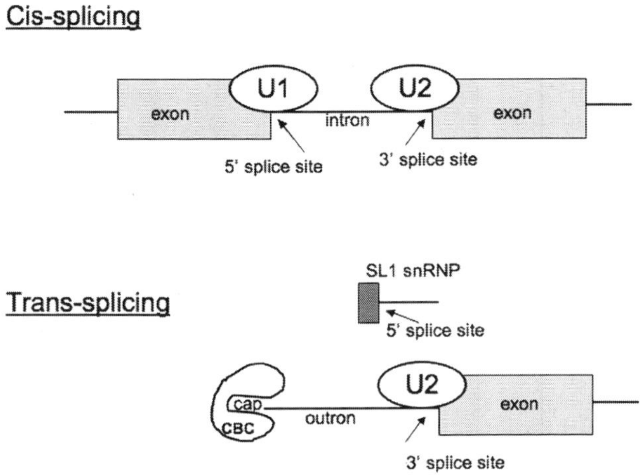

Figure 1.16. Various sorts of transcript splicing.

Source: Blumenthal T. Trans-splicing and operons in C. elegans. In: WormBook: The Online Review of C. elegans Biology [Internet]. Pasadena (CA): WormBook; 2005-2018. Figure 1, Comparison of cis- and trans-splicing. Available from: https://www.ncbi.nlm.nih.gov/books/NBK19704/figure/transsplicingoperons_figure1/

Mutations that trigger faulty splicing have been linked to a variety of human illnesses, such as cancer (Faustino and Cooper 2003). BRCA1; HGH, BRCA2, spinal muscular atrophy (SMA), cystic fibrosis, Wilms tumor suppressor, myotonic dystrophy (MD) linked with Frasier syndrome (WT1), as well as several other genes have been linked to faulty splicing and human disorders (Frith et al., 2006; Rossin et al., 2011).

In higher species, alternative splicing is the primary origin of protein abundance. By inserting and removing codons in the resultant mRNA, alternative splicing not only enhances the number of proteins and also changes the character of the protein. It can also alter the mRNA's reading frame. By inserting a termination codon into the mRNA, it might force protein production to stop. Modifications in regulatory components that impact mRNA stabilization and the translation procedure may regulate gene expression via alternative splicing.

Understanding the genome orders of mice or humans has indicated that alternative splicing has a significant impact on speciation. Both have a similar set of genes, and most of them have similar introns or exons. Furthermore, it's thought that about a quarter of the exons that experience alternative splicing are unique to humans and not seen in mice. Similarly, primates have alternately spliced exons which set the stage for primate evolution. The primate-specific exons appear to be derived from mobile genetic components with alu sequences. As a result, alu sequences are unique to primates.

1.8 PROTEIN CHEMISTRY BEFORE PROTEOMICS

Proteins have been called "natural robots" because they appear to know precisely what they need to do both inside and outside of cells (Tanford and Reynolds 2004). A protein's function, like that of several other substances, is governed by its structure. Proteins can serve a variety of purposes, as previously stated (shown in Table 1.1). Before proteomics science, most of the basic biology of proteins had been known (see Bell and Bell 1988 and Stryer 1982). This had been made feasible by the development of technologies for separating and purifying proteins, and also determining their particular action, the sequence and composition of amino acids, and three-dimensional structure. Methods for characterizing various physical and biological features, as well as their modulation and artificial production, had also been established (Subramanian & Kumar, 2004).

1.8.1 Separation and Purification of Proteins

During the production of the cellular extract, proteins were isolated from one another. Proteins can be extracted from tissue or cells using a variety of ways. Proteins have been segregated by precipitation in various concentrations of ammonium salts, which is usually done in a stepwise fashion. The molecular weights and charges of partly purified proteins have been utilized to segregate them. A tiny number of proteins is typically refined via ultracentrifugation inside a sucrose gradient depending upon differences in molecular weights. Gel filtration, which serves as a molecular sieve to segregate protein molecules dependent upon their sizes, is another form of separation. The molecular sieve sepharose (Sephadex; Sigma-Aldrich, St. Louis, MO) is routinely utilized to segregate protein molecules of various sizes. Ion-exchange chromatography can also segregate proteins depending upon their net +ve or -ve charges. In this type of ion-exchanger matrix, diethyl aminoethyl (DEAE) cellulose and carboxymethyl (CM) cellulose have been employed. Other chromatographic technologies that segregate protein molecules via their sizes and also their charges are generated. Apart from chromatography on a solid matrix, like sepharose, liquid chromatography and high-performance liquid chromatography (HPLC) are also generated. A wide range of proteins has been homogeneously purified. The 3D structures of some proteins are established after they are crystallized (Stoebel et al., 2008).

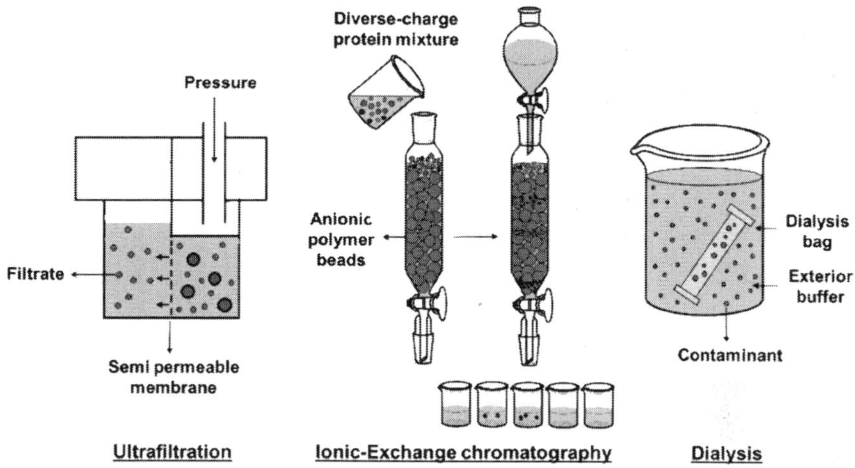

Figure 1.17. Protein separation represented graphically.

Source: Echave J, Fraga-Corral M, Garcia-Perez P, Popović-Djordjević J, H. Avdović E, Radulović M, Xiao J, A. Prieto M, Simal-Gandara J. Seaweed Protein Hydrolysates and Bioactive Peptides: Extraction, Purification, and Applications. Marine Drugs. 2021; 19(9):500. https://doi.org/10.3390/md19090500

In addition to various chromatographic techniques, many electrophoresis techniques are developed to segregate protein molecules depending upon their mass or charges on a conventional gel or a capillary gel by applying an electrical field. To segregate proteins of various molecular weights, sodium dodecyl sulfate (SDS) has been injected into the gel matrix. Because SDS is extremely -ve charged, all proteins in a mixture become similarly -ve charged when it is present. As a result, all proteins travel in an electrical field throughout electrophoresis in the existence of SDS depending upon its molecular sizes rather than their charges. During electrophoresis in an SDS-containing gel, smaller proteins move considerably quicker than larger proteins. Electrophoresis is a gel containing a combination of ampholine of various isoelectric points that may also be used to segregate a mixture of proteins depending upon net charges. These 2 electrophoresis procedures in SDS and ampholine gels are integrated whereas a protein combination is separated the 1st one in SDS gel and subsequently in ampholine gel depending upon electrical charges and molecular sizes. When 2 planes at right angles to one another are run in this approach, it segregates proteins depending upon charges or sizes. Proteins have been separated into distinct spots and colored with a dye throughout electrophoresis. More than 1100 *E. coli* proteins were separated sequentially on a two-dimensional gel for the 1st time. The capability of a two-dimensional gel to segregate an organism's whole protein composition and offer information about it in a single try ushered in a new era in the study of proteomics (Radford, 1991).

1.8.1.1 Specific Activity of Proteins

The preparation activity of a protein per milligram of that protein is known as its particular action. Enzymatic activity, capability to bind to a ligand, and biological activity are all used to evaluate a protein›s activity. A protein›s particular activity rises as its purity level rises. There have been various ways to figure out how much protein is in a generation. The most straightforward method is to measure light absorption at 280 nm. There are also various colorimetric techniques available, the most popular of which are the Bradford and Lowry systems (see Bell and Bell, 1988).

1.8.1.2 Determination of Molecular Weight

The molecular weight of a protein is an important necessity. It provides detail on the relative sizes of protein molecules. Chromatography with a matrix-like Sepharose or ultracentrifugation or the mobility of protein molecules in SDS gels on electrophoresis utilizing known protein markers are all common methods for determining molecular weight.

1.8.1.3 Amino Acid Composition

Proteins are made up of twenty distinct amino acids. The wide availability of such constituent amino acids in a protein is crucial. Knowing the number of various amino acids has also been useful for calculating the amino acid sequence in a protein molecule. A protein has been hydrolyzed in 6 M HCl for several hours to ascertain its amino acid content, and afterwards segregated via chromatography or electrophoresis. To make an electrophoretogram easier to see, the specific amino acid spots are colored using ninhydrin. Colorimetric analysis determines the number of amino acids within every location since the intensity of dye in every spot is proportional to the number of amino acids. Instead, the amino acids are segregated as eluents via chromatography being stained using a fluorescent dye, and the number of amino acids in each eluant is estimated spectroscopically, as the number of amino acids is dependent on the amount of dye taken by the amino acids. The entire procedure is mechanized, and an amino acid analyzer, a commercially available machine, can identify the amino acid makeup of a protein in a short time. Rockefeller University in New York City was the 1st to invent the amino acid analyzer (Harth et al., 1996).

1.8.1.4 Amino Acid Sequence

The N-terminus is used as a starting point for the sequential determination of the amino acid order in a protein. The Edman degradation reaction, which was initially invented at Rockefeller University and afterwards mechanized in Melbourne, Australia by Edman and his coworkers, is utilized to determine which amino acid is at the N-terminus (1950, 1967).

To classify a protein, it is often split up into smaller pieces called peptides which each contain about 50 amino acids. This can be done either through cyanogen bromide cleavage or through tryptic digestion. Firstly, the individual peptides have been isolated from each other. The peptide has been then adsorbed upon a solid surface, including a glass fiber that has been coated with the cationic polymer polybrene. PTH, also known as phenylisothiocyanate, is an Edman reagent that is applied to an adsorbed peptide that is suspended in a trimethylamine-based basic buffer solution. PTH interacts with an amino group of the N-terminal amino acid in this solution, and then that amino group is preferentially removed from the peptide by the injection of anhydrous acid. Isomerization of the amino acid with modified N-terminal results in the formation of phenylthiohydantoin. This is removed by washing, and subsequent chromatography allows for identification. After this, the cycle is repeated to identify the subsequent N-terminal amino acid in the peptide which is still adsorbed upon the glass fiber that has been coated using polybrene. The Edman degradation approach is sophisticated, but it has several drawbacks that must be considered. If the N-terminal amino acid is inhibited or buried in the majority of the protein, the approach would not operate.

1.8.2 Chemical Synthesis of Protein

Protein synthesis via chemical means has a long history. Emil Fischer was the first to synthesize the dipeptide glycylglycine in 1901. Subsequently, he created an octadecapeptide have a distinct amino acid order that included 15 glycine and 3 leucine residues. He couldn't regulate the amino acid order throughout the peptide synthesis. Zervas and Bergmann (1932) in Germany made a significant contribution in this regard by providing ways for preserving the amino group. Zervas and Bergmann both attended Rockefeller University in 1935, where they trained several protein biochemists, notably Standford Moore and William Stein, who had been awarded the Nobel Prize in 1972. In 1954, du Vigneaud et al. used Zervas and Bergmann technique to manufacture the octapeptide hormone oxytocin. The creation of oxytocin earned Vincent du Vigneaud the Nobel Prize in Chemistry in 1955. Such approaches for chemical synthesis in the solution medium, on either hand, had been time-consuming. Merrifield achieved a key breakthrough in protein chemistry by creating solid-phase synthesis at Rockefeller University in 1963. An amino acid is connected to impermeable support by its carboxyl end and then treated by another amino acid have an activated carboxyl group while being shielded by an alpha-amino group in this process. The dipeptide's amino group is then deprotected by removing the protective group at the amino-terminal, and a 3rd amino acid is reacted with a protected amino group and an activated carboxyl group to produce the tripeptide. Such a cyclic procedure of activation, deprotection, and protection is repeated until the whole protein or peptide has been synthesized. It is critical to preserve particular reactive amino acid side chains throughout the chemical synthesis. All protected groups have been deprotected at the end of the chemical production, and the peptide is subsequently split off the strong support. The biological and biochemical features of these proteins or peptides are next tested to confirm their identification with the naturally produced protein. This approach was employed by Merrifield to create the 1st enzyme named ribonuclease A. (RNaseA). The chemical synthesis procedure has now been mechanized. A machine designed at Rockefeller University performs the full peptide synthesis.

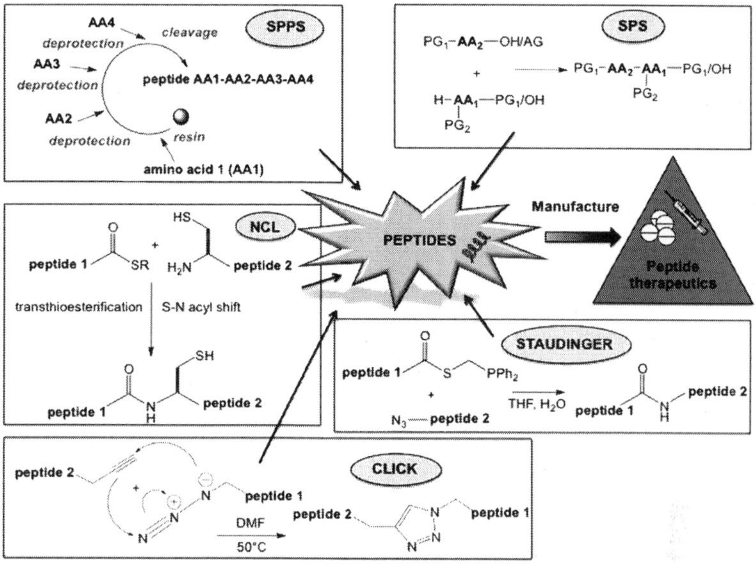

Figure 1.18. Chemical Synthesis of Proteins.

Source: Chandrudu S, Simerska P, Toth I. Chemical Methods for Peptide and Protein Production. Molecules. 2013; 18(4):4373-4388. https://doi.org/10.3390/molecules18044373

It's worth noting that the Merrifield group's chemical synthesis of RNaseA relied heavily upon their understanding of the amino acid order. In a cell, the process of protein biosynthesis often begins with the N-terminal amino acid, whereas the process of chemical synthesis typically begins with the C-terminal amino acid, which has been attached to an insoluble solid support. First, a succession of component peptides for a long-chain protein such as RNaseA is generated *in vitro*. These peptides are then ligated together to form the full-length protein. During the process of chemical synthesis, it is common practice to make use of either a base-sensitive 9-fluorenylmethyloxycarbonyl (Fmoc) group or an acid-sensitive tert-butoxycarbonyl (Boc) group to successfully secure the alpha-amino group of the amino acid that will be added to the growing chain of peptides. There are other places where a detailed method for the chemical production of proteins can be found (Nilsson et al. 2005).

1.8.3 Protein Engineering

Protein engineering technology arose from our capability to create proteins both *in vitro* and in vivo. Proteins of relevance having specific desirable features can be manufactured in large quantities using this approach. Protein engineering is carried out using 2 strategies that aren't even necessarily contradictory. Many laboratories produce proteins using both approaches. The 1st approach is known as "logical design." This necessitates a thorough understanding of the protein structure, which had been challenging before the introduction of proteomics and is now easily accessible. This procedure is cost-effective and utilizes site-directed mutagenesis. "Directed evolution" is the 2nd way. Since proteins of various types are created via random mutagenesis and afterwards chosen for desirable properties, this technique closely resembles natural evolution. Splicing DNA encoding various proteins to create an end-product, which combines the desirable properties of various proteins is sometimes done. The fundamental disadvantages of this approach are twofold: first, it is a time-consuming procedure that needs multiple constructions, and second, it needs high productivity, which is not attainable for some proteins.

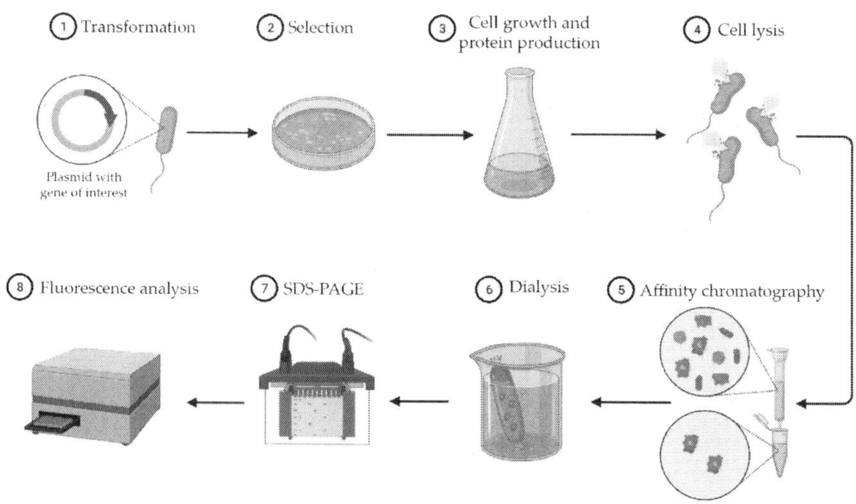

Figure 1.19. Protein Engineering Cycle.

Source: Gomes LC, Ferreira C, Mergulhão FJ. Implementation of a Practical Teaching Course on Protein Engineering. Biology. 2022; 11(3):387. https://doi.org/10.3390/biology11030387

1.8.4 Crystal Structure

The examination of the structure of the protein crystal reveals the three-dimensional structure of a protein in terms of the location of each atom in the amino acid chain of the protein one by one. The X-ray diffraction trend of the protein crystal typically presents the three-dimensional structure of the protein. Whenever an X-ray beam is shone on a protein crystal, the electrons in the atom disperse the X-ray, creating an X-ray diffraction pattern. To construct the three-dimensional structure, the X-ray pattern is next exposed to Fourier transformation evaluation. In the Cavendish laboratory in the 1940s, Max Perutz and John Kendrew investigated the three-dimensional protein. It required nearly twenty-two years for them to figure out the three-dimensional structure of myoglobin (Kendrew 1961) and hemoglobin (Perutz et al. 1960), for which both Perutz and Kendrew were awarded the Nobel Prize in 1962. Protein three-dimensional structural analysis developed slowly despite their work. By 1990, the X-ray diffraction pattern of protein crystals had disclosed the structure of lower than one hundred proteins.

The development of a novel method known as mad, wherein an X-ray beam had been directed at protein crystals using a synchrotron, was responsible for the overall acceleration of the procedure. The stage of diffraction could be easily determined using this method, which gave the data quickly. Both data on the phase and amplitude are needed to generate the structure in three dimensions. The X-ray diffraction of protein crystals that included heavy metals at distinct places, which necessitated the analysis of certain X-ray diffraction patterns, was used to identify an earlier stage. The increase in processing power expedited not only the process of analyzing three-dimensional structures but also the procedure overall. The three-dimensional structure of tiny proteins in solution can be determined using X-rays and nuclear magnetic resonance (NMR), among other techniques. NMR is useful for studying proteins that are unable to crystallize. In contrast to the three-dimensional structure of a protein crystal, the three-dimensional structure of a protein in its dynamic state can be determined using NMR. Protein molecules exist in the solution.

1.8.5 Regulation of Proteins and Active Site

Understanding the function of a protein is a crucial element of proteomics. This is especially important for knowing the function of protein in sickness and medication development. Proteins have numerous functions in a cell. Many proteins work by either catalyzing a biological reaction or binding to specific molecules, such as other proteins. An active site is a portion of a protein's structure; with the advancement of proteomics, the active site of an enzyme may now be identified utilizing bioinformatics software. During an enzymatic reaction, the active site binds to a substrate and then catalyzes the activity. The catalysis and binding of a substrate are explained using 2 models. The 1st is known as the "lock and key" model, while the 2nd is known as the "induced fit" model. The substrate and the active site have a lock and key link in the 1st model, which explains their uniqueness. The active site in the induced fit model isn't a hard structure, implying that the binding of a substrate causes some elasticity in the active site. Molecules that resemble the substrate's structure may bind to the active site and impede binding to the substrate; that's the foundation for inhibition of enzymes via medicines and enzyme modulation, as detailed below.

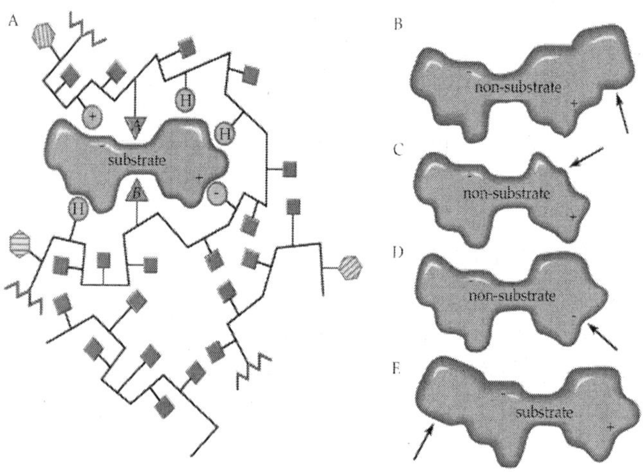

Figure 1.20. The active site's function in the lock-and-key fit of a substrate (the key) to an enzyme (the lock).

Source: https://cdn.britannica.com/13/6513-004-2CDF9DFD/role-site-fit-substrate-enzyme.jpg

In the case of non-enzyme proteins, these sites can attach to other proteins or specific compounds. Several medicines that are structurally identical to the substrate can bind to the active site of the protein and block its enzymatic activity by preventing it from interacting with the natural substrate. Since they contend with the substrate for the active site of the protein, these inhibitors are considered competitive inhibitors. Although chemicals attach to a protein's active site, some other molecules attach to the protein at a place beside the substrate-binding site, causing the protein's structure to change and preventing it from binding to the substrate. Since they don't compete with the substrate for attaching to the active site, these inhibitors are classified as noncompetitive inhibitors and allosteric inhibitors. Michaelis-Menton kinetics discriminates between non-competitive and competitive inhibitors. This type of kinetic study is done via plotting 1/V versus 1/S, wherein S is the concentration of substrate and V is the biological reaction velocity (see Bell and Bell 1988), resulting in a straight line. Monod and Jacob (1964) in Paris discovered the allosteric control of proteins. Monod and Jacob defined the function of an allosteric protein or a repressor protein implicated in the regulation of Lac operon transcription in this bacteria by creating suitable mutants. In 1965, both Monod and Jacob were awarded the Nobel Prize for such achievement.

1.8.6 Protein Targeting and Signal Sequence

After being synthesized on ribosomes, proteins travel throughout the cell to their final destinations, where they not only take up residence but also perform a variety of functions depending on the region of the cell in which they are found. In the 1970s, Gunter Blobel from Rockefeller University recognized approximately fifteen amino acid long patterns in various proteins. Such patterns target proteins to their locations, which include the cell membrane, the cell wall, and various organelles such as the Golgi bodies, peroxisomes, nucleus, chloroplasts, and mitochondria as well as the outer of the cell. Such Signal patterns function similarly to the zip codes that have been used to direct mail carriers to the correct addresses. The signal patterns are often found at the beginning of the protein, which is known as the N-terminus. After the proteins have been transported, the signal patterns are often broken by a protease. Proteins that are being sent to various locations typically carry signal patterns that comprise a distinct amino acid composition. For instance, proteins that are intended for the endoplasmic reticulum are composed of signal patterns that encompass five to ten hydrophobic amino acids at the N-terminus, while the signal pattern of proteins which are being transported to the nucleus comprises +ve charged amino acids inside the peptide. The pattern of amino acids that make up the mitochondrial targeting signals alternates between hydrophobic and positively charged amino acids. Typically, a signal pattern consisting of 3 amino acids located on the C-terminus of a protein is required for it to be targeted to peroxisomes. Such proteins are often in their unfolded state while being accompanied by a chaperon protein during their journey to their destination. Once the transport process is finished, the unfolded proteins are refolded with the assistance of a protein called a chaperon so that they can take their tertiary structures. Glycosylation, also known as the acquisition of the carbohydrate moiety, may oftentimes lead to the successful conclusion of the protein's journey.

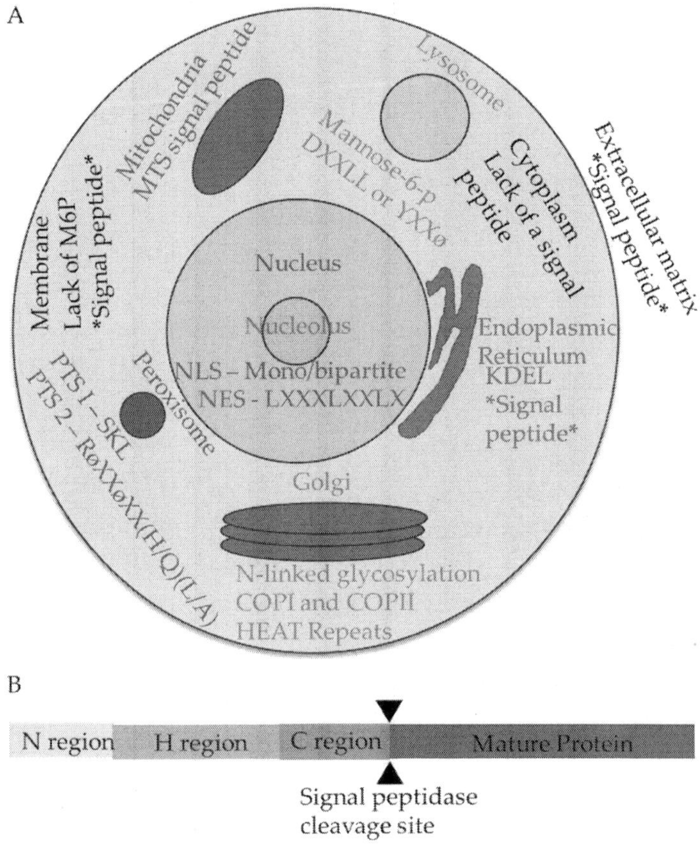

Figure 1.21. Protein targeting.

Source: Kellogg MK, Miller SC, Tikhonova EB, Karamyshev AL. SRPassing Co-translational Targeting: The Role of the Signal Recognition Particle in Protein Targeting and mRNA Protection. International Journal of Molecular Sciences. 2021; 22(12):6284. https://doi.org/10.3390/ijms22126284

Several human disorders are caused by a genetic deficiency in protein transport. As a result, studying protein transport is critical in gaining a better understanding of various human disorders and can give clues for treatment. Blobel won the Nobel Prize for discovering the process of protein transport in cells in 1999.

Over time, opinions on the distribution of enzymes and proteins within the cell have shifted. Previously, it had been assumed that enzymes were randomly dispersed in the cytosol and that enzymatic reactions occurred as a result of a chance encounter between a substrate molecule and an enzyme. In contrast to popular belief, enzymes from identical or related metabolic pathways have been clustered together in the cytosol rather than being scattered randomly. They occur together close to one another due to structural similarities that aid in their recognition.

1.8.7 Intein

Inteins are fragments of a protein that self-expel, succeeded by the exteins, which unite the existing fragments. The C-terminal and the N-terminal exteins have been connected by a peptide link as fast as the peptide is produced from the messenger RNA after the intein is removed. Inteins in proteins are similar to introns in genes in that they should be eliminated to create a functional protein, just as an intron in a transcript should be eliminated to create an interpretable message. Over two hundred inteins are identified in various proteins, and a database of inteins is accessible. The average length of a protein is between one hundred and eight hundred amino acids. Certain inteins are made up of 2 genes; for instance, cyanobacteria's dnae DNA polymerase (dnaE) comprises 2 elements, an N-intein segment with 123 amino acids and a C-intein segment with thirty-six amino acids. The 2 fragments have been encoded by 2 distinct genes for the alpha subunit of DNA polymerase III, dnaE-c and dnaE-n. In the situation of genes, this would be comparable to trans-splicing. The endonuclease found in gene-encoding inteins aids in the propagation of inteins. Inteins are discovered in eukaryotes, bacteria, and archaea among other forms of life. Protein engineering and protein marking for NMR characterisation are two examples of how proteins are employed. Inteins could be a beneficial tool in the development of a medicine that prevents the loss of intein from a protein, rendering it nonfunctional and hence accountable for the sickness.

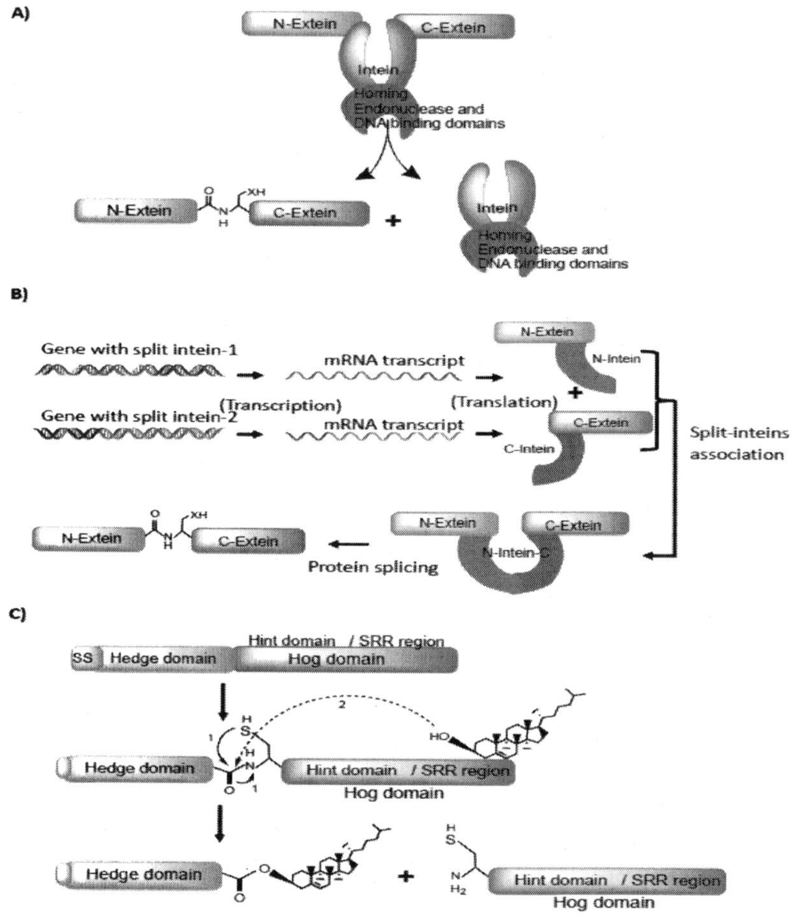

Figure 1.22. A summary of Intein.

Source: Pavankumar TL. Inteins: Localized Distribution, Gene Regulation, and Protein Engineering for Biological Applications. Microorganisms. 2018; 6(1):19. https://doi.org/10.3390/microorganisms6010019

1.9 UNSTRUCTURED PROTEIN

It has been demonstrated that there have been 2 types of proteins found in biological systems: one type of protein has a structure that has been arranged, while the other type of protein is fundamentally disordered and unorganized (Dyson and Wright 2005).

Before the advent of proteomics, a good deal of research had been done to figure out the structure of the 1st set of proteins. It is well knowledge that proteins can take several different levels of the organization, including tertiary, secondary, and primary structures. The quaternary structure is an additional degree of protein organization that is present in several proteins. The fundamental structure of a protein is made up of the order in which its amino acids are arranged. Since the folding in the main structure is dependent upon the interconnections of the amino acids, especially their side-chain amongst themselves, the secondary structure shows the coiled structure of the protein. It's because the main structure folds into itself. A protein's three-dimensional structure, known as its tertiary structure, is determined by the polypeptide chain's ability to completely fold back on itself. The function of the protein and its three-dimensional shape are both determined by its tertiary structure. After taking a tertiary structure, several proteins bind with one another or with another protein to assume a quaternary structure. Proteins that have the same peptides serving as subunits in their quaternary structures are referred to as homomers, while proteins that have distinct peptides serving as subunits in their quaternary structures are referred to as heteromers.Hemoglobin is a great instance of a protein that has a quaternary structure since it consists of four different chains: 2 intermediate chains, 2 beta chains, and 2 alpha chains. As long as the proteins have been created, there have been no proteins that reside in their main structure like a straight stretch of peptide because of the instant biochemical interactions that occur among the various amino acids that have been present in the stretch of the polypeptide.

Circular dichroism or even gel filtration may be used to assess the secondary structure; however, this is not always the case. There are a few various approaches that may be taken to uncover the various protein structures. Not every protein has a three-dimensional structure. Because they do not possess a tertiary structure, it is believed that more than thirty-five percent of the proteins that may be identified in living systems do not have an inherent structure. These proteins have been given the name "intrinsically unstructured proteins" (Dyson and Wright 2005). The fundamental structures of such proteins typically do not contain any of the bulky amino acids that are hydrophobic. Such proteins only have a short lifespan and exist as chains of random coils. In most cases, they are responsible for carrying out regulatory duties, like directing the regulation of the cell cycle, controlling translation and transcription, and signaling passageways. Although research into such proteins is just getting started, it is anticipated that it will provide a great deal of light on how the functions of proteins are controlled.

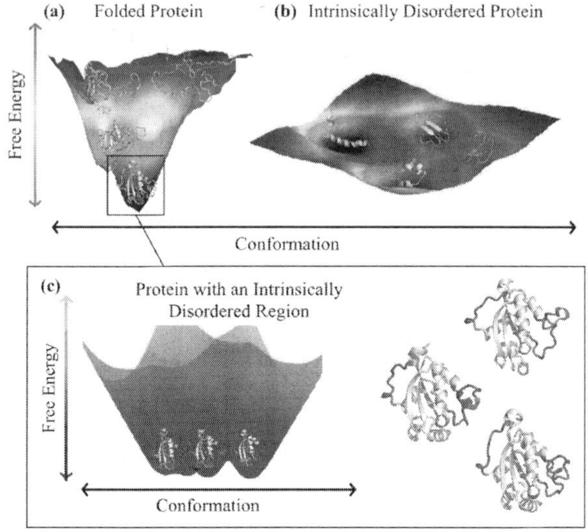

Figure 1.23. Induced folding of intrinsically disordered proteins.

Source: Burger VM, Gurry T, Stultz CM. Intrinsically Disordered Proteins: Where Computation Meets Experiment. Polymers. 2014; 6(10):2684-2719. https://doi.org/10.3390/polym6102684

1.10 PROTEIN MISFOLDING AND HUMAN DISEASE

The random coil shape of the intrinsically unorganized proteins, such an indeterminate form can be assumed by other proteins as a result of mutation or other modifications in the proteins. The structure of proteins is changed in several disorders, including neurodegenerative disorders and other illnesses. They are devoid of the tertiary and secondary structures that are often found in proteins. In most cases, they are the consequence of proteins failing to fold correctly. Anfinsen demonstrated that the main structure of the protein is responsible for controlling the tertiary and secondary structures of the protein. He was able to demonstrate that denaturing circumstances, like the injection of urea into the solution that contained the protein, caused the enzyme ribonuclease to undergo a process that led to its unfolding. Ribonuclease that hasn't folded properly is inactive as an enzyme. Furthermore, once urea is withdrawn from the protein solution via dialysis, ribonuclease begins to fold and eventually acquires a fully folded form with a tertiary and secondary structure as well as the full recovery of its enzymatic activity. Because of this achievement, Anfinsen was awarded the Nobel Prize in 1972. Moreover, because they have been manufactured within the cell, several proteins are unable to fold because they remain unfolded. The presence of proteins that have not folded correctly can lead to several illnesses in human beings. The misfolding of proteins is the root cause of several neurodegenerative disorders, like Creutzfeldt-Jakob disease, Alzheimer's disease, Kuru, and mad cow disease.

Certain proteins which induce mad cow disease had been identified as infectious protein complexes in the 1960s. Prions, which are similar to virions, had been named as they only included proteins; the nucleic acid comprised the infectious particles. Prusiner received the Nobel Prize in Physiology or Medicine in 1997 for its work defining prions as protease-resistant proteins (PrPs). After a mutation in the gene for regular PrP, prions are formed wherein the mutant PrP may not fold appropriately. Such misfolded PrPs had been later discovered to be infectious protein molecules that spread by triggering the misfolding of proteins that may have exited as fully functional, perfectly folded proteins. Misfolding of a protein known as the cystic fibrosis transmembrane conductance regulator (CFTR) causes various disorders, including cystic fibrosis. The biological activity essential for the transportation of chloride ions has been lost when misfolded proteins stay in the random coil state. Consequently, with Alzheimer's disease, they become persistent and form the distinctive beta-sheet plaques in the sufferers' brains.

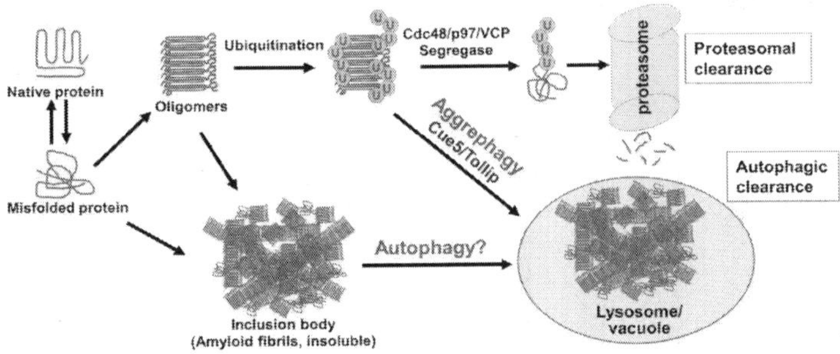

Figure 1.24. Protein Misfolding.

Source: Folger A, Wang Y. The Cytotoxicity and Clearance of Mutant Huntingtin and Other Misfolded Proteins. Cells. 2021; 10(11):2835. https://doi.org/10.3390/cells10112835

A protein molecule is normally destroyed when it becomes misfolded for any cause. Furthermore, it can sometimes survive disintegration and subsequently function as a chaperon, causing other protein molecules to misfold. The so-called infectivity of prions, which induces Creutzfeldt-Jakob disease and mad cow disease, is based on this. When a healthy cow is administered unfolded protein or prions, it induces other proteins in the brain cells to misfold, resulting in illness.

When prions had been originally found, they had been believed to be infected protein particles, and it had been believed for a time that they worked as a proteinaceous infectious agent in the same way that infective viral DNAs/RNAs did. This viewpoint contradicted the long-held belief that only nucleic acids function as genetic information. The idea of a protein as genetic information is resolved due to the discovery that the production of novel prion particles is caused by the misfolding of existing naturally present proteins. Because prions are unable to replicate, they don't code for daughter prions. Rather than, these prions may cause misfolding of newly synthesized proteins encoded by the host genome, which attracts more prion particles. In cystic fibrosis, for instance, a misfolded CFTR protein causes additional CFTR proteins to misfold, and the cell loses its normal function. It has been discovered that normal PrP regulates long-term memory in mammals. Prions, or prion-like particles, are discovered in yeast and fungus, such as Podospora, wherein they influence various phenotypes in the organisms that contain them. Therefore, understanding appropriate protein folding is critical to understanding the causes of various disorders and how to cure them. The yeast heat shock protein Hsp40/YdjI has been recognized as a protein that prevents misfolded proteins from aggregating and aids in their refolding. It appears to detect specific repeat patterns as a protein's consensus motif. DnaJ, an *E. coli* protein, is similar to yeast protein. This protein could help researchers better understand human disorders involving protein misfolding.

REFERENCES

1. AjitTamadaddi, C., &Sahi, C. (2016). J domain independent functions of J proteins. *Cell Stress and Chaperones*, *21*(4), 563-570.
2. Akam, M. E., Roberts, D. B., Richards, G. P., &Ashburner, M. (1978). Drosophila: the genetics of two major larval proteins. *Cell*, *13*(2), 215-225.
3. Alberts, B., Johnson, A., Lewis, J., Raff, M., Roberts, K., & Walter, P. (2002). The shape and structure of proteins. In *Molecular Biology of the Cell. 4th edition*. Garland Science.
4. Alsagaby, S. A. (2019). Understanding the fundamentals of proteomics. *Curr. Top. Pept. Protein Res*, *20*(3), 25.
5. Andersen, S. E., Bulman, L. M., Steiert, B., Faris, R., & Weber, M. M. (2021). Got mutants? How advances in chlamydial genetics have furthered the study of effector proteins. *Pathogens and disease*, *79*(2), ftaa078.
6. Anfinsen C. B. 1973 Principles that govern the folding of protein chains. Science 181, 223.
7. Arruda, S. C. C., de Sousa Barbosa, H., Azevedo, R. A., &Arruda, M. A. Z. (2011). Two-dimensional difference gel electrophoresis applied for analytical proteomics: fundamentals and applications to the study of plant proteomics. *Analyst*, *136*(20), 4119-4126.
8. Aschaffenburg, R. (1968). Section G. Genetics. Genetic variants of milk proteins: their breed distribution. *Journal of Dairy Research*, *35*(3), 447-460.
9. Avery, O. T., C. M. MacLeod, and M. McCarthy. 1944. Studies on the chemical nature of the substance inducing transformation of pneumococcal types I induction of transformation by DNA from Pneumococcus type III. J. Exp. Med. 79, 137
10. Axe, D. D. (2004). Estimating the prevalence of protein sequences adopting functional enzyme folds. *Journal of molecular biology*, *341*(5), 1295-1315.
11. Ayala, F. J. and J. A. Kiger. 1984. Modern Genetics, Second Edition. Menlo Park, CA: Benjamin Cummings.
12. Baker, C. A., Manwell, C., Labisky, R. F., & Harper, J. A. (1966). Molecular genetics of avian proteins—V. Egg, blood and tissue proteins of the ring-necked pheasant, *Phasianus colchicus* L. *Comparative Biochemistry and Physiology*, *17*(2), 467-499.

13. Barisoni, L., Schnaper, H. W., & Kopp, J. B. (2009). Advances in the biology and genetics of the podocytopathies: implications for diagnosis and therapy. *Archives of pathology & laboratory medicine*, *133*(2), 201-216.

14. Barlow, C. K., & O'Hair, R. A. (2008). Gas-phase peptide fragmentation: how understanding the fundamentals provides a springboard to developing new chemistry and novel proteomic tools. *Journal of mass spectrometry*, *43*(10), 1301-1319.

15. Beadle, G. W. and E. L. Tatum. 1941. Genetic Control of Biochemical Reactions in Neurospora. Proc. Nat. Acad. Sci. U. S. A. 27 (11), 499–506.

16. Begun, D. J., Whitley, P., Todd, B. L., Waldrip-Dail, H. M., & Clark, A. G. (2000). Molecular population genetics of male accessory gland proteins in Drosophila. *Genetics*, *156*(4), 1879-1888.

17. Bell, J. E. and E. T. Bell. 1988. Proteins and Enzymes. Englewood Cliffs, NJ: Prentice Hall.

18. Bergmann, M. and Zervas L. 1932. U¨ bereinallgemeinesVerfahren der Peptid-Synthese.*BerichtederDeutschenChemischenGesellschaft*65(7), 1192–1201.

19. Berzelius, J. J. 1838. The word "protein" was coined from Greek word proteios meaning the first by J¨onsJakob Berzelius in 1838 in a letter to his friend.

20. Biron, D. G., Hughes, A. L., Loxdale, H. D., &Moura, H. (2007). The need for megatechnologies: massive sequencing, proteomics and bioinformatics. *Encyclopedia of Infectious Diseases: Modern Technologies*, 357-377.

21. Bishop, A., Buzko, O., Heyeck-Dumas, S., Jung, I., Kraybill, B., Liu, Y., ... &Shokat, K. M. (2000). Unnatural ligands for engineered proteins: new tools for chemical genetics. *Annual review of biophysics and biomolecular structure*, *29*(1), 577-606.

22. Bludau, I., &Aebersold, R. (2020). Proteomic and interactomic insights into the molecular basis of cell functional diversity. *Nature Reviews Molecular Cell Biology*, *21*(6), 327-340.

23. Bonetta, R., & Valentino, G. (2020). Machine learning techniques for protein function prediction. *Proteins: Structure, Function, and Bioinformatics*, *88*(3), 397-413.

24. Bousquet-Dubouch, M. P., Fabre, B., Monsarrat, B., &Burlet-Schiltz, O. (2011). Proteomics to study the diversity and dynamics of proteasome complexes: from fundamentals to the clinic. *Expert review of proteomics*, *8*(4), 459-481.

25. Brenner, S., F. Jacob, and M. Meselson. 1961. An unstable intermediate carrying information from genes to ribosomes for protein synthesis. Nature 190, 576–581.

26. Caestecker, M. D., & Meyrick, B. (2001). Bone morphogenetic proteins, genetics and the pathophysiology of primary pulmonary hypertension. *Respiratory research*, *2*(4), 1-5.

27. Chaijaroenkul, W., Thiengsusuk, A., Rungsihirunrat, K., Ward, S. A., & Na-Bangchang, K. (2014). Proteomics analysis of antimalarial targets of *Garcinia mangostana* Linn. *Asian Pacific Journal of Tropical Biomedicine*, *4*(7), 515-519.

28. Chemale, G., Morphew, R., Moxon, J. V., Morassuti, A. L., LaCourse, E. J., Barrett, J., ... &Brophy, P. M. (2006). Proteomic analysis of glutathione transferases from the liver fluke parasite, *Fasciola hepatica*. *Proteomics*, *6*(23), 6263-6273.

29. Chen, F., Cseke, L. J., Lin, H., Kirakosyan, A., Yuan, J. S., & Kaufman, P. B. (2016). The study of plant natural product biosynthesis in the pregenomics and genomics eras. *Natural Products From Plants. Taylor & Francis*, 203-220.

30. Chiti, F., Calamai, M., Taddei, N., Stefani, M., Ramponi, G., & Dobson, C. M. (2002). Studies of the aggregation of mutant proteins *in vitro* provide insights into the genetics of amyloid diseases. *Proceedings of the National Academy of Sciences*, *99*(suppl 4), 16419-16426.

31. Choi, B. H., Park, G. T., & Rho, H. M. (1999). Interaction of hepatitis B viral X protein and CCAAT/enhancer-binding protein α synergistically activates the hepatitis B viral enhancer II/pregenomic promoter. *Journal of Biological Chemistry*, *274*(5), 2858-2865.

32. Colozza, M., de Azambuja, E., Cardoso, F., Bernard, C., &Piccart, M. J. (2006). Breast cancer: achievements in adjuvant systemic therapies in the pre-genomic era. *The Oncologist*, *11*(2), 111-125.

33. Colozza, M., De Azambuja, E., Personeni, N., Lebrun, F., Piccart, M. J., & Cardoso, F. (2007). Achievements in systemic therapies in the pregenomic era in metastatic breast cancer. *The oncologist*, *12*(3), 253-270.

34. Crick, F. H. 1970. Central dogma of molecular biology. Nature 227, 561–563.

35. Crick, F. H. C. 1958. Biosynthesis of macromolecules. Symp. Soc. Exp. Biol. XII, 138–163.

36. Darwin, C. 1859. On the Origin of Species, 1st ed. London, UK: John Murray.

37. D'HOOGHE, T. M., Kyama, C., Debrock, S., Meuleman, C., &Mwenda, J. M. (2004). Future directions in endometriosis research. *Annals of the New York Academy of Sciences*, *1034*(1), 316-325.

38. du Vigneaud, V., C. Ressler, J. M. Swan, C. W. Roberts, and P. G. Katsoyannis. 1954. Oxytocin: synthesis. *J. Am. Chem. Soc.* 76; 3115–3118.

39. Dyson, H. J. and P. E. Wright. 2005. Elucidation of the protein folding landscape by NMR. Methods Enzymol. 394, 299–321.

40. Ecker, A., Bushell, E. S., Tewari, R., &Sinden, R. E. (2008). Reverse genetics screen identifies six proteins important for malaria development in the mosquito. *Molecular microbiology*, *70*(1), 209-220.

41. Edman, P. 1950. Method for determination of the amino acid sequence in peptides. ActaChemica Scandinavia. 4, 283–284.

42. Edman, P. and G. Begg. 1967. A protein Sequenator Eur. J. Biochem. 1, 80–91.

43. Emilsson, V., Ilkov, M., Lamb, J. R., Finkel, N., Gudmundsson, E. F., Pitts, R., ... &Gudnason, V. (2018). Co-regulatory networks of human serum proteins link genetics to disease. *Science*, *361*(6404), 769-773.

44. Faustino, N. A. and T. A. Cooper. 2003. Pre-mRNA splicing and human disease. Genes Dev 17, 419–437.

45. Ferré, J., Escriche, B., Bel, Y., & Van Rie, J. (1995). Biochemistry and genetics of insect resistance to *Bacillus thuringiensis* insecticidal crystal proteins. *FEMS Microbiology Letters*, *132*(1-2), 1-7.

46. Fire, S. Q., M. K. Xu, S. A. Montgomery, S. E. Kostas, and C. C. Driver. 1998. Potent and specific genetic interference by double-stranded RNA in *Caenorhabditis elegans*. Nature. 391, 806–811

47. Fischer, E. 1902. *Nobel Lectures, Chemistry 1901–1921* , Amsterdam, The Netherlands: Elsevier 1966

48. Florens, L., Washburn, M. P., Raine, J. D., Anthony, R. M., Grainger, M., Haynes, J. D., ... &Carucci, D. J. (2002). A proteomic view of the

Plasmodium falciparum life cycle. *Nature, 419*(6906), 520-526.

49. Floros, J., & Hoover, R. R. (1998). Genetics of the hydrophilic surfactant proteins A and D. *Biochimica et BiophysicaActa (BBA)-Molecular Basis of Disease, 1408*(2-3), 312-322.
50. Frith, M. C., Forrest, A. R., Nourbakhsh, E., Pang, K. C., Kai, C., Kawai, J., ... &PLoS Genetics EIC Wayne Frankel. (2006). The abundance of short proteins in the mammalian proteome. *PLoS genetics*, *2*(4), e52.
51. Garattini, E., Mendel, R., Romão, M. J., Wright, R., &Terao, M. (2003). Mammalian molybdo-flavoenzymes, an expanding family of proteins: structure, genetics, regulation, function and pathophysiology. *Biochemical Journal*, *372*(1), 15-32.
52. Gorrod, A. E. 1909. Inborn Errors of Metabolism. Oxford, UK: Oxford University Press.
53. Harth, G., Lee, B. Y., Wang, J., Clemens, D. L., &Horwitz, M. A. (1996). Novel insights into the genetics, biochemistry, and immunocytochemistry of the 30-kilodalton major extracellular protein of Mycobacterium tuberculosis. *Infection and immunity*, *64*(8), 3038-3047.
54. Hershey, A. D. and M. Chase. 1952. Independent functions of viral protein and nucleic acids in growth of bacteriophage. J. Gen. Physiol. 36, 39–56
55. Hinman, M. N., & Lou, H. (2008). Diverse molecular functions of Hu proteins. *Cellular and molecular life sciences*, *65*(20), 3168-3181.
56. Hu, J. C. (2000). A guided tour in protein interaction space: coiled coils from the yeast proteome. *Proceedings of the National Academy of Sciences*, *97*(24), 12935-12936.
57. Huberts, D. H., & van der Klei, I. J. (2010). Moonlighting proteins: an intriguing mode of multitasking. *Biochimica et BiophysicaActa (BBA)-Molecular Cell Research*, *1803*(4), 520-525.
58. Ingram, V. M. 1956. A specific chemical difference between globins of normal and sickle-cell an´umiahemoglobins. Nature 178, 792–794.
59. Ingram, V. M. 1957. Gene mutations in human hemoglobin: the chemical difference between normal and sickle h´umoglobin. Nature 180, 326–328.
60. Jacob, F., and J. Monod. 1964. Biochemical and genetic mechanisms of regulation in the bacterial cell. Bull. Soc. Chim. Biol. 46, 1499–1532.

61. Jiang, Y., Oron, T. R., Clark, W. T., Bankapur, A. R., D'Andrea, D., Lepore, R., ... &Radivojac, P. (2016). An expanded evaluation of protein function prediction methods shows an improvement in accuracy. *Genome biology*, *17*(1), 1-19.
62. Jurkiewicz, A., Caricati-Neto, A., &Jurkiewicz, N. H. (2009). Functionomics: the analysis of a postgenomic concept on the basis of pregenomic pharmacological studies in smooth muscle. *Anais da Academia Brasileira de Ciências*, *81*(3), 605-613.
63. Kendrew, J. 1961. The three-dimensional structure of a protein molecule. Sci. Am. 205, 96–110.
64. Khorana, H. G. 1968 Nucleic acid synthesis in the study of Genetic code. Nobel Lecture 341–366.
65. Klose, J. 1975. Protein mapping by combined isoelectric focusing and electrophoresis in mouse tissues. A novel approach to testing for induced point mutations in mammals. Humangenetik 26: 231–243.
66. Kniemeyer, O. (2011). Proteomics of eukaryotic microorganisms: The medically and biotechnologically important fungal genus Aspergillus. *Proteomics*, *11*(15), 3232-3243.
67. Kuhn, M., & Goebel, W. (2006). Genomics of Listeria monocytogenes. *Pathogenomics: Genome Analysis of Pathogenic Microbes*, 339-366.
68. Lander, E. et al. 2001. International Human Genome Sequencing Consortium: Initial sequencing and analysis of the human genome. Nature 409, 860–921.
69. Leder, P., and M. W. Nirenberg. 1964. RNA codewords and protein synthesis3.On the nucleotide sequence of a cysteine and leucine RNA code words. Proc. Na. Acad. Sci. U.S.A. 52, 1521–1529.
70. Lee, D., Redfern, O., &Orengo, C. (2007). Predicting protein function from sequence and structure. *Nature reviews molecular cell biology*, *8*(12), 995-1005.
71. Lee, H., Lee, Y. H., Huh, Y. S., Moon, H., & Yun, Y. (1995). X-gene Product Antagonizes the p53-mediated Inhibition of Hepatitis B Virus Replication through Regulation of the Pregenomic/Core Promoter (*). *Journal of Biological Chemistry*, *270*(52), 31405-31412.
72. Lewin, B. 2004. Gene VIII. Upper Saddle River, NJ: Prentice Hall.
73. Li, Y., Ito, M., Sun, S., Chida, T., Nakashima, K., & Suzuki, T. (2016). LUC7L3/CROP inhibits replication of hepatitis B virus via suppressing

enhancer II/basal core promoter activity. *Scientific reports*, *6*(1), 1-11.

74. Lill, R. (2009). Function and biogenesis of iron–sulphur proteins. *Nature*, *460*(7257), 831-838.
75. Loscalzo, J. (2011). Systems biology and personalized medicine: a network approach to human disease. *Proceedings of the American Thoracic Society*, *8*(2), 196-198.
76. Maestri, E., Klueva, N., Perrotta, C., Gulli, M., Nguyen, H. T., &Marmiroli, N. (2002). Molecular genetics of heat tolerance and heat shock proteins in cereals. *Plant molecular biology*, *48*(5), 667-681.
77. Makjaroen, J., Somparn, P., Hodge, K., Poomipak, W., Hirankarn, N., &Pisitkun, T. (2018). Comprehensive proteomics identification of IFN-λ3-regulated antiviral proteins in HBV-transfected cells. *Molecular & Cellular Proteomics*, *17*(11), 2197-2215.
78. Marschalek, R. (2011). Mechanisms of leukemogenesis by MLL fusion proteins. *British journal of haematology*, *152*(2), 141-154.
79. Matheson, N. J., Sumner, J., Wals, K., Rapiteanu, R., Weekes, M. P., Vigan, R., ... &Lehner, P. J. (2015). Cell surface proteomic map of HIV infection reveals antagonism of amino acid metabolism by Vpu and Nef. *Cell host & microbe*, *18*(4), 409-423.
80. Mattick, J. S. 2003. Challenging the dogma: the hidden layer of non-protein-coding RNAs in complex organisms. BioEssays 25, 930–939.
81. Mendel, J. G. 1866. Versuche ¨uberPlflanzenhybridenVerhandlungen des naturforschendenVereines in Br¨unn, Bd. IV f¨ur das Jahr, 1865. Abhandlungen, 3–47. For the English translation, see: Druery, C. T. and W. Bateson. 1901. Experiments in plant hybridization. J. R. Horticul. Soc. 26, 1–32.
82. Merrifield, R. B. 1963. Solid Phase Peptide Synthesis. I. The Synthesis of a Tetrapeptide. J. Am. ChemSci J. 85(14), 2149–2154.
83. Mishra, N. C. 2002. Nucleases—Molecular Biology and Applications. New York: Wiley.
84. Mitchell, H. K. and J. Lein. 1948. A Neurospora mutant deficient in the enzymatic synthesis of tryptophan. J. Biol. Chem. 175, 481–482.
85. Mitchell, H. K., M. B. Houlahan, J. Lein. 1948. Some aspects of genetic control of tryptophan metabolism in Neurospora. Genetics 33, 620.
86. Moore, S. and W. Stein. 1972. The chemical synthesis of pancreatic ribonuclease and deoxyribonuclease. Nobel Lecture 80–93.

87. Muthuirulan, P. (2016). Insight into Phenotypic and Genotypic Discrimination of Bacterial Pathogens: From Pre-Genomic to Post-Genomic Era. *BAOJ Med Nursing*, *2*, 018.

88. Nabieva, E., Jim, K., Agarwal, A., Chazelle, B., & Singh, M. (2005). Whole-proteome prediction of protein function via graph-theoretic analysis of interaction maps. *Bioinformatics*, *21*(suppl_1), i302-i310.

89. Nasrallah, M. E., Barber, J. T., & Wallace, D. H. (1970). Self-incompatibility proteins in plants: detection, genetics, and possible mode of action. *Heredity*, *25*(Pt. 1), 23-27.

90. Neidhardt, F. C., VanBogelen, R. A., & Vaughn, V. (1984). The genetics and regulation of heat-shock proteins. *Annual review of genetics*, *18*(1), 295-329.

91. Neuwald, A. F., Aravind, L., Spouge, J. L., &Koonin, E. V. (1999). AAA+: A class of chaperone-like ATPases associated with the assembly, operation, and disassembly of protein complexes. *Genome research*, *9*(1), 27-43

92. Nilsson, B. L., M. B. Soellner, and R. T. Raines. 2005. Chemical synthesis of proteins. Annu. Rev. Biophys. Biomol. Struct. 34, 91–118.

93. O'Farrell, P. H. 1975. High resolution two-dimensional electrophoresis of proteins. J. Biol. Chem. 250, 4007–4021.

94. Paape, D., &Aebischer, T. (2011). Contribution of proteomics of *Leishmania* spp. to the understanding of differentiation, drug resistance mechanisms, vaccine and drug development. *Journal of proteomics*, *74*(9), 1614-1624.

95. Pauling Linus 1954 in Nobel Lectures, *Chemistry 1942-1962* , Elsevier Publishing Company, Amsterdam, 1964.

96. Payne, P. I. (1987). Genetics of wheat storage proteins and the effect of allelic variation on bread-making quality. *Annual Review of Plant Physiology*, *38*(1), 141-153.

97. Payne, P. I., Holt, L. M., Jackson, E. A., & Law, C. N. (1984). Wheat storage proteins: their genetics and their potential for manipulation by plant breeding. *Philosophical Transactions of the Royal Society of London. B, Biological Sciences*, *304*(1120), 359-371.

98. Payne, P. I., Holt, L. M., Lawrence, G. J., & Law, C. N. (1982). The genetics of gliadin and glutenin, the major storage proteins of the wheat endosperm. *Plant Foods for Human Nutrition*, *31*(3), 229-241.

99. Pellegrini, A. (2003). Antimicrobial peptides from food proteins. *Current pharmaceutical design*, *9*(16), 1225-1238.

100. Perutz, M. F. Rossman, MG; Cullis, AF; Muirhead, H; Will, G; North, 1960. Structure of haemoglobin: a three-dimensional Fourier synthesis at 5.5A resolution, obtained by x-ray analysis. Nature 185, 416–422.

101. Prucca, C. G., I. Salvin, R. Quiroga, E. V. Elias, F. D. Rivero, A. Saura, P. G. Carranza, and H. D. Lujan. 2008 Antigenic variation in Giardia lamblia is regulated by RNA interference. Nature 456, 750–754.

102. Radford, A. (1991). Methods in yeast genetics—A laboratory course manual by M Rose, F Winston and P Hieter. pp 198. Cold Spring Harbor Laboratory Press, Cold Spring Harbor, New York. 1990. $34 ISBN 0-87969-354-1.

103. Rasheed, A., Xia, X., Yan, Y., Appels, R., Mahmood, T., & He, Z. (2014). Wheat seed storage proteins: Advances in molecular genetics, diversity and breeding applications. *Journal of Cereal Science*, *60*(1), 11-24.

104. Rittmann, B. E., Krajmalnik-Brown, R., &Halden, R. U. (2008). Pre-genomic, genomic and post-genomic study of microbial communities involved in bioenergy. *Nature reviews microbiology*, *6*(8), 604-612.

105. Rossin, E. J., Lage, K., Raychaudhuri, S., Xavier, R. J., Tatar, D., Benita, Y., ... & Daly, M. J. (2011). Proteins encoded in genomic regions associated with immune-mediated disease physically interact and suggest underlying biology. *PLoS genetics*, *7*(1), e1001273.

106. Sanger, F. 1952. The arrangement of amino acids in proteins. Adv. Protein Chem. 7, 1–28

107. Sanger, F. 1958. The chemistry of insulin. Nobel Lecture 544–556.

108. Sarabhai, A. S., A. W. O. Stretton, S. Brenner, and A. Bolle. 1964. Colinearity of the gene with the peptide chain. Nature 201, 13–17.

109. Schlessinger, A., Punta, M., &Rost, B. (2007). Natively unstructured regions in proteins identified from contact predictions. *Bioinformatics*, *23*(18), 2376-2384.

110. Setlow, P. (1988). Small, acid-soluble spore proteins of Bacillus species: structure, synthesis, genetics, function, and degradation. *Annual Reviews in Microbiology*, *42*(1), 319-338.

111. Sharp, P. A. 2005. The discovery of split genes and RNA splicing. Trends Biochem Sci. 30, 279–81.

112. Shewry, P. R., Halford, N. G., &Lafiandra, D. (2003). Genetics of wheat gluten proteins. *Advances in genetics*, *49*, 111-184.

113. Shishkin, S. S., Kovalyov, L. I., &Kovalyova, M. A. (2004). Proteomic studies of human and other vertebrate muscle proteins. *Biochemistry (Moscow)*, *69*(11), 1283-1298.

114. Snipes, G. J., & Suter, U. E. L. I. (1995). Molecular anatomy and genetics of myelin proteins in the peripheral nervous system. *Journal of anatomy*, *186*(Pt 3), 483.

115. Song, J., & Singh, M. (2009). How and when should interactome-derived clusters be used to predict functional modules and protein function?. *Bioinformatics*, *25*(23), 3143-3150.

116. Stoebel, D. M., Dean, A. M., &Dykhuizen, D. E. (2008). The cost of expression of *Escherichia coli* lac operon proteins is in the process, not in the products. *Genetics*, *178*(3), 1653-1660.

117. Stryer, L. 1982. Biochemistry, 2nd edition. San Francisco, CA: W.H. Freeman Co.

118. Subramanian, S., & Kumar, S. (2004). Gene expression intensity shapes evolutionary rates of the proteins encoded by the vertebrate genome. *Genetics*, *168*(1), 373-381.

119. Sumner, J. B. 1946. The chemical nature of enzyme. Nobel Lectures 114–121

120. Sun, S., Nakashima, K., Ito, M., Li, Y., Chida, T., Takahashi, H., ... & Suzuki, T. (2017). Involvement of PUF60 in transcriptional and post-transcriptional regulation of hepatitis B virus pregenomic RNA expression. *Scientific reports*, *7*(1), 1-15.

121. Sunagar, K., Morgenstern, D., Reitzel, A. M., & Moran, Y. (2016). Ecological venomics: How genomics, transcriptomics and proteomics can shed new light on the ecology and evolution of venom. *Journal of Proteomics*, *135*, 62-72.

122. Swanson, H. I., & Bradfield, C. A. (1993). The AH-receptor: genetics, structure and function. *Pharmacogenetics*, *3*(5), 213-230.

123. Taha, T. Y., Anirudhan, V., Limothai, U., Loeb, D. D., Petukhov, P. A., & McLachlan, A. (2020). Modulation of hepatitis B virus pregenomic RNA stability and splicing by histone deacetylase 5 enhances viral biosynthesis. *PLoS pathogens*, *16*(8), e1008802.

124. Tanford, C. and J. Reynolds. 2004. Nature's Robot: A History of Proteins. Oxford, UK: Oxford University Press.

125. Thomas, P. D., Campbell, M. J., Kejariwal, A., Mi, H., Karlak, B., Daverman, R., ... &Narechania, A. (2003). PANTHER: a library of protein families and subfamilies indexed by function. *Genome research*, *13*(9), 2129-2141.

126. Vandahl, B. B., Birkelund, S., & Christiansen, G. (2004). Genome and proteome analysis of Chlamydia. *Proteomics*, *4*(10), 2831-2842.

127. Virgin 4th, H. W., Mann, M. A., Fields, B. N., & Tyler, K. L. (1991). Monoclonal antibodies to reovirus reveal structure/function relationships between capsid proteins and genetics of susceptibility to antibody action. *Journal of virology*, *65*(12), 6772-6781.

128. Volkin, E. and L. Astrachan. 1957. Phosphorus incorporation in *Escherichia coli* ribo-nucleic acid after infection with bacteriophage T2. *Virology* 1956,149–161.

129. Watson, J. 1965. Molecular Biology of Gene. Melno Park, CA: W. A. Benjamin.

130. Watson, J. D. and F. H. C. Crick 1953a. Molecular structure of nucleic acids: A structure for desoxyribonucleic acids. Nature 171 731

131. Watson, J. D. and F. H. C. Crick 1953b. General implications of the structure of desoxyribonucleic acids. Nature 171. 964

132. Watson, J. D., Laskowski, R. A., & Thornton, J. M. (2005). Predicting protein function from sequence and structural data. *Current opinion in structural biology*, *15*(3), 275-284.

133. Wilkins, M. 1996. 1997. Protein identification in the post-genome era: the rapid rise of proteomics. Q. Rev. Biophys. 30(4), 279–331.

134. Xia, X. (2018). Fundamentals of Proteomics. In *Bioinformatics and the Cell* (pp. 421-436). Springer, Cham.

135. Yanofsky, C. 1952. The effect of gene changes on tryptophan desmolase formation. Proc. Nal. Acad. Sci. U.S.A. 38, 215–226.

136. Yanofsky, C. 2005a. The favorable features of Tryptophan synthetase for proving Beadle and Tatum's one gene—one enzyme hypotheis. *Genetics* 169, 511–516.

137. Yanofsky, C. 2005b. Using studies on Tryptophan metabolism to answer basic biological questions. *J. Biol. Chem.* 278, 19859–10878.

138. Yanofsky, C., B. C. Carlton, J. R. Guest, D. R. Helinski, and U. Henning. 1964. On the colinearity of gene structure and protein structure. *Proc. Nat. Acad. Sci.* U.S.A. 51, 266–27

139. Yao, Z., Jia, X., Megger, D. A., Chen, J., Liu, Y., Li, J., ... & Yuan, Z. (2018). Label-free proteomic analysis of exosomes secreted from THP-1-derived macrophages treated with IFN-α identifies antiviral proteins enriched in exosomes. *Journal of proteome research*, *18*(3), 855-864.

140. Zhang, J., Keene, C. D., Pan, C., Montine, K. S., &Montine, T. J. (2008). Proteomics of human neurodegenerative diseases. *Journal of Neuropathology & Experimental Neurology*, *67*(10), 923-932.

CHAPTER 2

PROTEOMICS—RELATION TO GENOMICS

CONTENTS

2.1 INTRODUCTION

Proteomics is the research among all proteins as well as their relationships in a cell or a species; as a result, proteomics is inextricably linked to genomics, which is the study of all genetics in just about any species as well as how genes translate for proteins. Bioinformatics, also known as computational biology, aids in the interpretation as well as administration of all data produced during the investigation of an organism's genomes and proteome. Since the origin of genes or the DNA pattern stays similar in all cells in the human body, genomics is static. Proteomics, on the other hand, is dynamic because protein patterns vary from one kind of cell to the next or throughout developmental stages. The several phases of an insect's life highlight the striking difference between biology and genetics: The caterpillar and the butterfly, for instance, have the same DNA, yet their proteomes or protein profiles are separate, giving them opposite forms and content as if they were two separate species (Figure 2.1).

Roughly 30,000 genes in humans are essential for the manufacture of 4.5 million proteins. It is difficult to understand how a tiny set of genes can produce just too many proteins. Variant splicing of transcripts and posttranslational alteration of proteins are responsible for most of the variation in protein composition (Zivy& de Vienne, 2000; Martyniuk&Denslow, 2009).

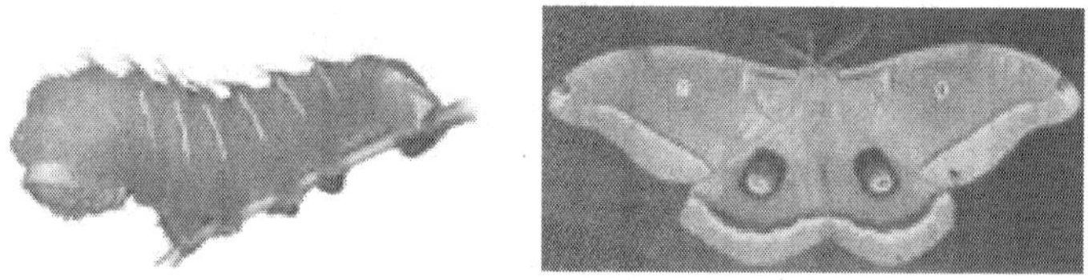

Figure 2.1. The caterpillar and butterfly exemplify the differences in proteomics at two different stages in the life cycle of an insect.

Source: https://www.charismaticplanet.com/life-cycle-butterfly/

2.2 GENOMICS

Genomics involves determining the whole pattern of DNA of an individual's chromosome(s) and comprehending the structure of these patterns. And over 500 creatures' whole DNA patterns have been discovered. It started with the sequencing of a bacterial virus 174 as well as advanced to the human genome (see Table 2.1: Genomes of various organisms).

Table 2.1. Genomes of different organism

Organism	Year	Size	Comment
$\phi \times 174$ virus	1977	3 Kb	First bacterial virus, not a free-living organism sequenced.
Hemophilus influenze	1995	1.8 Mb	First bacterium or a free-living organism sequenced.
Saccharomyces cerevisiae	1996	12 Mb	Yeast, the first eukaryote sequenced
Caenorhabditis elegans	1998	100 Mb	Soil worm
Drosophila melanogaster	2000	180 Mb	Fruit fly
Human	2000	3000 Mb	The first draft of human genome sequence announced.
Chimpanzee	2005	3000 Mb	The closet living relative of human.
Neanderthal	2006	3000 Mb	The extinct cousin of human, more than a million base pairs sequenced.

A minimum of three types of DNA appear to occur in higher species. These DNA patterns are subsequently translated into proteins which regulate the shape and groups of organisms. The second category contains just translated DNAs; through the transmission of RNA (tRNA) and ribosomal RNA (rRNA) production, these RNAs facilitate the synthesis of other regions into proteins. The third type of DNA pattern, like a regulator, activator, promoter, and silencer sections, regulates the transcription of genes involved throughout transcription into tiny nuclear RNA (snRNA) as well as microRNA, or as whole DNA. Consequently, the third type of DNA pattern solely functions as a regulatory pattern to govern the transcription of other DNA patterns.

Moreover, certain DNA sequences may act as structural elements or other types of codes outside certain necessary for encoding proteins (Lan et al., 2003; Low et al., 2013). These DNA patterns are necessary for chromosomal stability and functioning. For instance, the nucleotide patterns in telomeres need not encode for any protein yet are necessary for the preservation of chromosomal size or integrity during replication. Similarly, the DNA sequences that make up the centromeres need not encode any proteins yet are necessary for the separation of daughter chromosomes at the time of cell division. Without correct centromere activity, cell division would lead to uneven chromosomes' division to daughter cells, a condition known as aneuploidy. Aneuploidy in people is responsible for multiple disorders, like Down syndrome, and illnesses, like cancer. Moreover, certain DNA sequences serve as regulatory areas or structural parts of the chromosome in undefined manners. The remaining 30–40% of DNA sequences are made up of retroposons collected during the evolution of species, such as humans. Consequently, the origin of a substantial fraction of the nucleotide sequences in larger species remains unclear. These DNA sequences are referred to as "junk DNA" since their purpose in the form and composition of the genome is unknown.

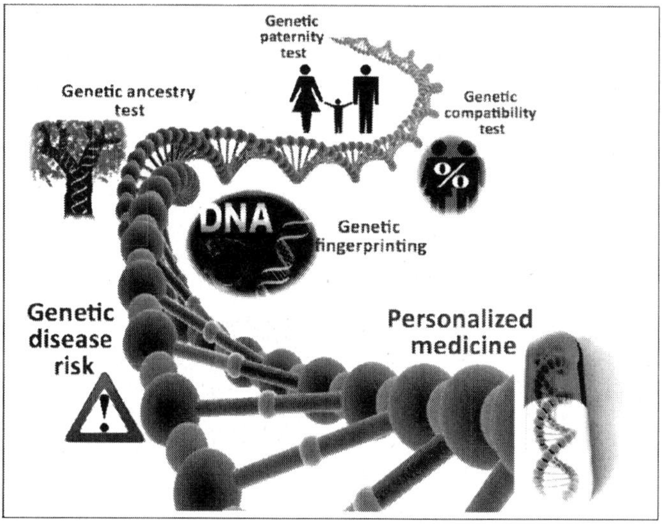

Figure 2.2. Graphical representation of genomics and its applications.

Source:https://databricks.com/glossary/genomics

Supposedly, genomics began with Watson and Crick's 1953 discovery of the double-helix structure of DNA. This DNA framework involves the characteristics of the genetic data. The property is such that the DNA strand is equivalent to the other inside a DNA molecule. Utilizing one strand as a material for the formation of the strand inside the DNA, the double helix offers the method for the duplication of genetic data. It also facilitates the storage and processing of genetic data. The data is contained as triplets or codons, which are composed of the four nucleotides (adenine, cytosine, guanine, and thymine) present in DNA. Data is relayed from DNA to RNA Molecule (mRNA) and ultimately to proteins by transcriptional regulation, respectively. Lastly, the DNA structure allows for mutagenesis or hereditary alteration in the genetic data in the event of a duplicating mistake that occurs in the duplication of the DNA molecule (Fabian et al., 2008; Oldham, 2009).

Till the identification of enzymes like DNA polymerases, constraint endonucleases, and ligases for the manufacture, cutting, and joining of DNA sections, genomics stayed a pipe dream. The advent of tools like DNA cloning, Gene sequencing (Maxam and Gilbert 1977, Sanger Nicklen and Coulson 1977), and DNA replication hastened the march to genomics (Mullis and Faloona 1987). The mechanization of large-scale DNA designing (Smith et al., 1985) enabled the sequencing of over one million nucleotides per day. The introduction of technology and accompanying software made it feasible to automate DNA sequencing. Computers also created an opportunity to organize and evaluate Genomic DNA data, paving the way for the initiation of genome projects (Collins and Galas 1993).

2.2.1 Human Genome Project and Other Genome Projects

After the technology for assembling DNA fragments had become accessible, it formed critical to producing projects to understand the DNA sequence of people and other species to better grasp how DNA needs to carry data for the design of living species into the adult stage, including a fertilized egg cell and their interrelatedness. To accomplish this, efforts were undertaken to create and enhance DNA sequencing techniques based on throughput and efficiency, as well as to undertake genome projects. All components of DNA sequencing were carried out concurrently. The Human Genome Project was developed in 1990 by the US Department of Energy and the National Institutes of Health to discover the whole Sequence of DNA of people in terms of understanding its function in human development and health. Many genetic creatures› genome studies were also conducted as experimental models for studying the significance of genes that can not be investigated in humans due to technological and ethical constraints. The noted the following of the human DNA sequence was disclosed and released in June 2000, and it was christened the language of God in which our destiny is written by W. Clin- ton, then-President of the United States. Whenever the human genome project was completed, the whole genomes of various model species were determined. Hemophilus, yeast, fruit fliesor *Drosophila melanogaster*, the roundworm *Caenorhabditis elegans*, and *Arabidopsis thaliana*, a model plant were among them (Mustafa, 2005; Gstaiger &Aebersold, 2009).

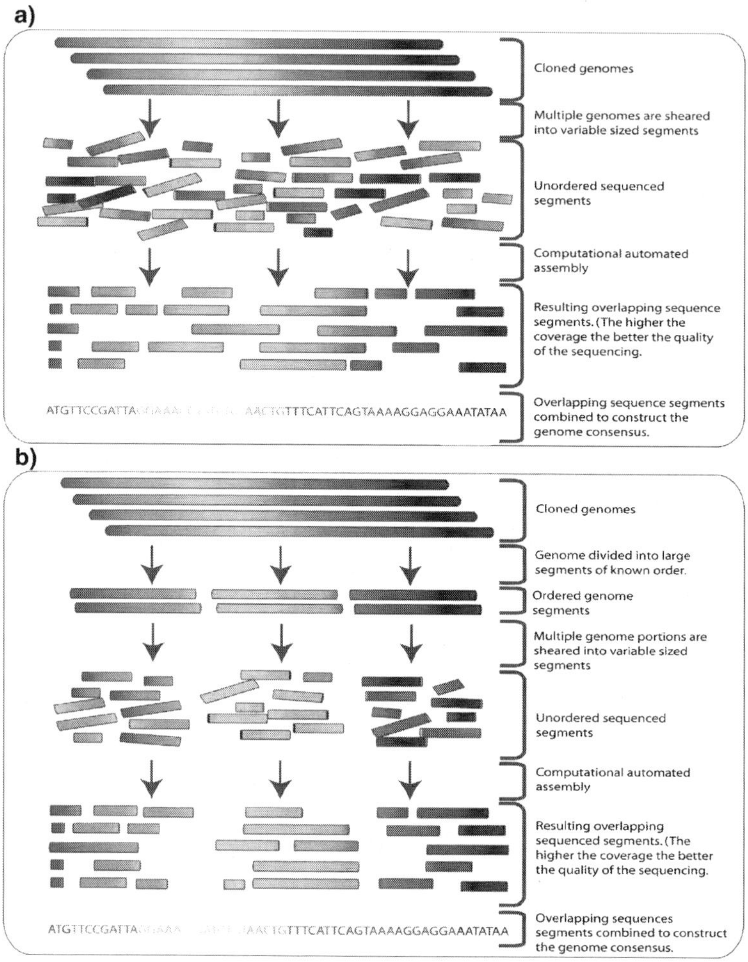

Figure 2.3. This illustration depicts the shotgun reading technique, which involves copying, breaking, scanning, and computer analysis of DNA to determine the initial genetic sequence.

Source: https://commons.wikimedia.org/wiki/File:Whole_genome_shotgun_sequencing_versus_Hierarchical_shotgun_sequencing.png

The DNA sequence of Hemophilus revealed that a single organism may survive with as few as 460 protein-coding genes. The DNA sequence of yeast revealed that a eukaryotic cell requires roughly 6000 genes. The number of genes necessary for distinct cellular metabolic processes was also found by analyzing the yeast genome. The fruit fly's DNA sequence revealed the existence of around 13,000 genes essential for the ongoing production of a multicellular creature (Kenyon et al., 2002; Gupta et al., 2016). Around 19,000 genes were discovered in the roundworm's genome investigation. This research also revealed the importance of programmed cell death in the worm's growth. The human genome investigation revealed the astonishing finding that just roughly 23,000 genes are necessary for the ongoing production of a complex creature from one fertilized cell to an adult comprising 1013 cells with over 700 cell types grouped into distinct tissues and organs. Originally, it was believed that humans had up to 100,000 genes. This figure was predicated on the presumption that each protein had one gene, however, it comes out that the human genome has less than 23,000 genes but also more than 100,000 proteins, which are made possible by alternative splicing of transcripts from a limited number of genes. The human genome project not just identified the whole patterns of around 3 billion nucleotides spread over 24 chromosomes, comprising 22 autosomes and X and Y chromosomes, but also released the technologies used throughout such procedure public. For the first time, our effort attempted to comprehend the ethical, legal, and social implications (ELSI) of the results. This endeavor was a significant divergence from previous scientific endeavors. The care with which ethical concerns were developed was unlike anything else before attempted. For instance, whenever the Manhattan Project was launched to produce a nuclear weapon, ethical issues were completely disregarded, which atomic physicists subsequently condemned (Yates, 2000; Wright et al., 2010).

2.2.2 Methods to Study the Genome Project

The invention and production of vectors, as well as the creation of recombinant DNA technology, were crucial in molecular cloning. By treating with restriction enzymes, human DNA and perhaps other DNA might be easily produced as required length pieces. These pieces might then be put into round vector DNA, like plasmid DNA, using ligases after being linearized by restriction endonucleases cut at a specific spot (Skylas et al., 2005; Özdemir et al., 2017).

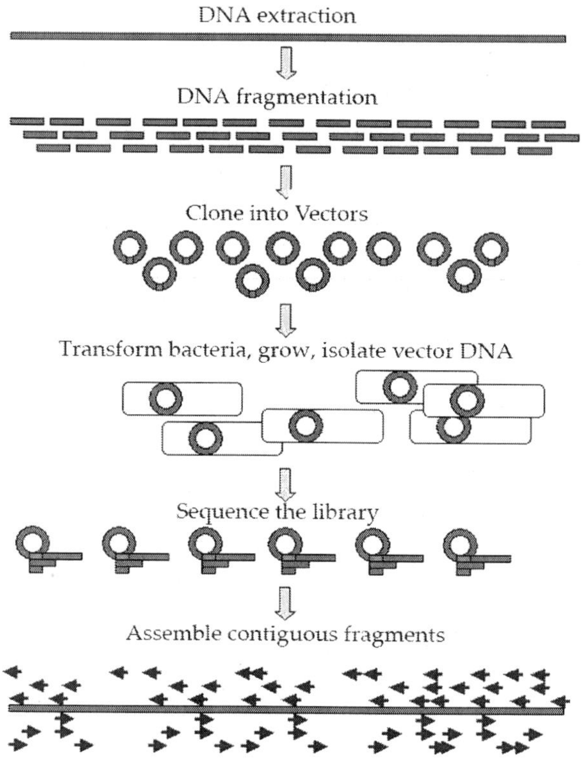

Figure 2.4. Clone and sequencing cover ranges.

Source: https://commons.wikimedia.org/wiki/File:DNA_Sequencing_gDNA_libraries.jpg

Transfection might be used to deliver recombinant vector DNA carrying mammalian or foreign DNA sequences as inserted into bacteria organisms like *Escherichia coli* (see Figure 2.2). Because recombinant vectors may self-replicate in bacterial host cells, they are promoted in those cells. Several approaches exist for dramatically increasing the number of chimerical vectors in the host cells.

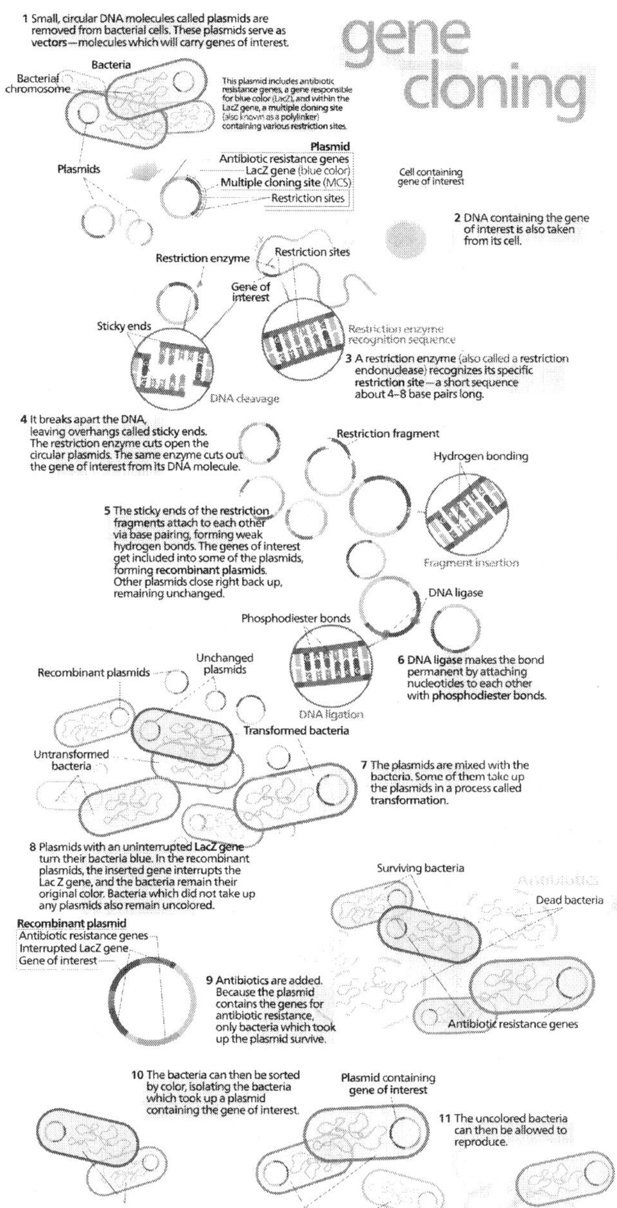

Figure 2.5. A general method for the cloning of a gene.

Source: By Kelvinsong - Own work, CC BY-SA 3.0, https://commons.wikimedia.org/w/index.php?curid=23730988

Many more types of vectors are now accessible in addition to plasmid DNA (Sali et al., 2003; McLean, 2013). Cosmid, phage DNA, and YAC (yeast artificial chromosome)), BAC (bacterial artificial chromosome), and PAC (bacterial artificial chromosome) (phage artificial chromosome) are examples of these. These vectors are excellent for cloning huge Genetic material and, as a result, are important in genome sequencing. Much genetic information of importance, like the one that causes human disorders, was cloned to use the recently invented DNA cloning technology detailed here before the commencement of genome sequencing. Random DNA sections first were cloned and afterward, bacterial cells with the cloned specific genes of relevance were chosen in this strategy to clone human genes or genes of relevance from many other resources. The shotgun technique of cloning was named after it (Pardanani et al., 2002; Baak et al., 2003). A group of bacterial colonies comprising an organism's whole genome is typically formed. The genomic library is a kind of gathering like this. Other types of DNA libraries exist in extra to the genomic collection created by the shotgun technique. After then, the cloned DNA sections from all these collections are utilized for analysis. Such DNA sections are sometimes expanded *in vitro* using the polymerase chain reaction (PCR) technology, which was invented by Mullis in 1985. (see Mullis &Faloona 1987). Mullis received the Nobel Prize for inventing the PCR technique. The computers finally organize the DNA sequences into the nucleotide sequence of a chromosome (Cox & Mann, 2007; Li et al ., 2014).

Following the establishment of a genomic library, more specialized techniques for cloning DNA from a particular chromosome or a specific tissue containing a specific gene were created. As a result, many DNA libraries were accessible. The genomic library, chromosomal collection, and expression collection are the three. A genomic library consists of bacterial cells that contain a whole genetic code achieved by the shotgun technique of cloning. A chromosome library consists of DNA sections from a specific chromosome contained in bacterial cells. The expression catalog, on the other hand, includes a total of demonstrated genes/DNA sections from a specific tissue, like blood cells, brain cells, or liver cells. The expression collection is made up of mRNA from a specific cell type. The mRNAs are extracted from a specific tissue as well as backward translated into complementary DNA (cDNA) using reverse transcription enzymes. Throughout the production of the expression collection, the single-stranded cDNA is made double using DNA polymerase and afterward put into vector DNA for proliferation in bacterial host cells (Kovac et al., 2013; Song & Lin, 2017).

Clones of either the protein-encoding or protein-noncoding portions of DNA from a species are found in the genomic and chromosomal libraries. Nevertheless, inserting them into chromosomes for the aim of tracing and synthesizing the whole Sequence of DNA chromosomes takes a lot of time and work. The tracing is a little simpler using the chromosome-specific library. It gives you the sequence of DNA of a chromosome from one side to another. For important information related, overlapping DNA fragments from a chromosome are beneficial. It is feasible to traverse from one side of the chromosome to another using overlapping areas (Bendixen, 2005; Rao &Swamy, 2008).

To build a chromosomal library, an individual's chromosomes should be isolated from one another and acquired in their original state. Varied species have various numbers and sizes of chromosomes. Lower eukaryotes, like yeasts, have significantly smaller chromosomes, containing only around 15 million strands or fewer. Pulsed-field gel electrophoresis is used to isolate them. Humans or other higher-order creatures have substantially bigger chromosomes. In general, human chromosomes are 150 million nucleotide bases long. The individual chromosome 1 has around 260 million base pairs, whereas the Y chromosome only has roughly 60 million. They couldn't be isolated till the cell sorter was invented in the late 1980s at the Los Alamos National Laboratory.

2.2.3 The Outcome of the Study of the Human Genome

There are few human genes, and gene identification is difficult. The small estimate of protein-encoding genes was among the major discoveries of the human genome. Originally, it was considered that humans could have up to 100,000 genomes since they could have up to 100,000 activities, most of which are regulated through one enzyme. This viewpoint was congruent with Beadle and Tatum's (1941) one-gene–one-enzyme approach, and it remained until the mRNA sequence for mRNA research got accessible. An mRNA array study revealed the existence of around 100,000 mRNAs. Nevertheless, as the human genome project was completed and the nucleotide sequences for people were accessible, it had become clear that only roughly 23,000 gene encoding elements were present in people (Cahill et al., 2001; Günther et al., 2014).

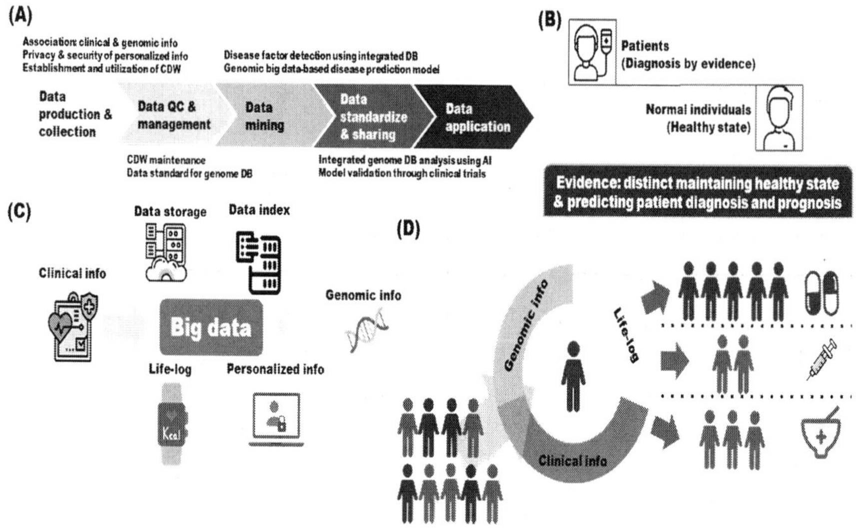

Figure 2.6. Overview of genome.

Source: Gim J-A. A Genomic Information Management System for Maintaining Healthy Genomic States and Application of Genomic Big Data in Clinical Research. International Journal of Molecular Sciences. 2022; 23(11):5963. https://doi.org/10.3390/ijms23115963

By gene splicing, a huge amount of mRNAs and proteins were produced in humans from a limited set of genes. It is presently believed that far more than half of the genetic mutations experience alternative splicing, results as in the estimation of three proteins per gene. On the appearance, it is incredible that humans (23,000 genes) have just a several thousand extra genes than a fruit fly (13,000 genes) or a roundworm (19,000 genes). However, several thousand genes might create a big impact and contribute to the emergence of a sophisticated creature like a human. This viewpoint becomes more understandable whenever it is realized as humans and chimps vary by several hundred genes, with fewer than ten genes accounting for morphological characteristics and a max of forty genes accounting for variations in intellectual growth (Fermin et al., 2006; Asgari &Mofrad, 2015).

The human genome research discovered the design of DNA that does not correspond to every protein. These non-protein-coding patterns account for about 95% of all human nucleotide patterns. These DNA sequences are not normal genes. They are referred to as junk DNA since they do not impact human features. Nevertheless, as such junk DNA is not rubbish since it has been preserved throughout the evolutionary course of human history as well as a change in it may halt human growth (Pognan, 2004; Latonen et al., 2018).

These strands are waste DNA similarly to that we collect rubbish in our basement in the hope that we could use it eventually. They are not junked DNA; if that was the case, this DNA would have been removed throughout the evolution change in the identical manner that we dispose of our rubbish. DNA hybridization tests by Britten and Kohne provided strong evidence for the occurrence of repeat sequences in higher creatures, like humans (1968).

The human genome project revealed the presence and placement of these repetitive or junk DNA sections on the chromosomes. And over half of these repetitive sequences include recurrent nucleotides like GCGCGC. These have been copied numerous times and scattered haphazardly across the genome between those protein-coding regions. Such segments with many duplicates emerge as a fast renaturing segment of the genome in DNA hybridization tests. Protein coding patterns, on the other hand, seem as a progressively renaturing segment of the genome since they are retained as a small chunk of the DNA pattern in the genome. Several such sequences provide light on our evolutionary history. A small region of roughly 300 base pairs known as an Alu sequence is found amongst the junk DNA. Alu sequences account for approximately 7% of the genetic code (i.e., much more than protein-coding patterns, which show only 5 percent of the genome). So the junk DNA, which contains Alu sequences, has been dubbed "selfish DNA." They are self-centered in the idea that their only goal is to survive as a component of the genome. Through the evolution change, these patterns can not be deleted. The Alu patterns are exclusive to primates (Benesch et al., 2007; Carpi et al., 2010).

Understanding their function in humans as well as other higher species is a huge problem. Many junk DNAs are thought to be RNA transcribing sequences that play functions in the procedure of translation or the regulation of gene activity or gene suppression. Additional junk DNAs may have a range of structural functions in linking genes, defining gene spacing, or managing chromosomal supercoiling as well as total chromosome integrity (Solé, & Pastor-Satorras, 2002).

Before genome sequencing, it was considered that all rational human beings had the same nucleotide patterns in their chromosomes. It was also thought that people who were affected by an illness varied solely in the DNA pattern of the gene(s) associated with producing the sickness. Nevertheless, after the full genome sequences of numerous normal people were decoded, it had become clear that their nucleotide sequences diverged in one or more locations. This variation in nucleotide sequences across normal people was referred to as an SNP. As a result, one person may have a sequence of AAGCCTA in a certain gene, but another person might also have AAGCTTA in a similar sequence in that gene, indicating an SNP. As a result, these two people show two distinct alleles of such a gene (i.e., the C allele and the T allele). SNPs are often found in populations where 1% or more of the people vary in their DNA sequence (Zivanovic et al., 2009). It is vital to note that SNPs are specific to a majority's ethnicity or geographic area. SNPs might arise in a gene's coding or non-coding sequences, as well as in intergenic areas between genes. Because of genetic code degeneracy, SNPs in a gene's coding sequence might nonetheless code for the similar amino acid in the protein. SNPs that result in the creation of similar proteins are referred to as alike, while SNPs that result in the creation of distinct proteins are referred to as nonsynonymous. SNPs in noncoding or intergenic areas may induce splicing or transcription factor binding errors, or they might modify the type of non-coding or roles In the regulation (Ju et al., 2010).

SNPs may be found all across the genetic code. They occur once per 300 nucleotide lengths of DNA. The human genome has around 10 million SNPs and 3 billion nucleotides. Cytosine is substituted by thymine in 2/3rd of all SNPs. Due to variations in the nucleotide sequence recognized by a restriction enzyme, SNPs are frequently discovered by their RFLP. Microarray examination or DNA sequencing is more effective in detecting SNPs. SNPs are important in the development of customized medicine as they determine an individual's reaction to a medication, chemical, or infection.

Each chromosome has an identical number of genes or DNA sequences. Deletion is the removal of a chromosomal section, while duplication is the increase in the amount of a certain section; both are the fundamental reasons for various human disorders or syndromes. The findings of the human genome, on the other hand, have shown the existence of a varied proportion of similar DNA segments, either with or in the absence of adverse consequences for the person. The other noteworthy discovery from the human genome project is a difference in the genome size of the length of DNA in the different chromosomes. It has been discovered that a length of DNA may occur in several copies on the sister chromatids of a similar person or in dissimilar people, with or without creating health issues.

This is referred to as CNV. The variance generated by differing DNA section sequence similarity considerably outnumbers the latter referred to by SNPs.

CNV has been demonstrated to alter gene transcription as well as HIV and malaria resistance. CNV has been linked to immunological problems and neoplasia, as well as complex illnesses including diabetes and heart disease. CNV may also influence an individual's capacity to adapt to a specific environment. And over 1400 CNVs have been discovered in the human genome. CNVs are found at several places on human as well as Chimpanzee chromosomes. Some hotspots are shared by both species, indicating that they have historical relevance. Surprisingly, CNVs for specific DNA sequences have been discovered in 80 percent of twin pairs. CNVs like this might be the cause of phenotypic variances between conjoined twins. CNV has been discovered in the DNA of people suffering from neuroblastoma, autism, and Alzheimer's disease.

Around 38 thousand human transcribed components have been discovered during the latest study of human cDNA as well as transcribed sequence tags. These human transcriptional units were compared to those discovered in other primates plus non-primate sequences, and 131 primate-specific transcriptional units were discovered. Approximately half of such transcriptional units conserved by primates include protein-coding sequences. These were likewise discovered to be devoid of introns, suggesting that they came from transposons. According to the expression profile of such primate-specific genes, they are solely expressed in the brain and reproductive systems. Around 21 of these primate-specific genes have been discovered to be included in the formation of human sperm, with several of them causing sterility in humans, such as teratozoospermia. The conclusion is that these primate-specific genes affected primate evolution. In several respects, humans are distinct from other animals and primates, notably monkeys, our nearest surviving family. This contains our capability to walk upright and bipedally, as well as our capability to speak, language development, and several other sophisticated characteristics that form the foundation of our civilization and culture, such as agriculture, architecture, gourmet preparing food, songs, watercolor, advanced weaponry systems, the goal is to find and create things, and a desire to understand the fundamentals of nature. So, what distinguishes us as human beings? This is an age-old conundrum. New findings in molecular biology and genomics, such as genome comparisons between humans and chimpanzees, appear to give a little insight into this subject.

Rapid alterations in primate genes are thought to be implicated in critical networks affecting sound perception, nerve signal propagation, cellular ions transport, and sperm generation, according to genomic research. Primates are distinguished from mammals and other species by the fast alterations in these gene groupings. The genetic basis for bipedal movement, a bigger brain, speech, and advanced language abilities are among the fundamental differences that separate humans from chimpanzees. Excluding the development of the capacity to speak, the genetic basis for such a task has as yet to be identified at the molecular level. It appears that a gene on chromosome 7 encoding a transcription factor protein called FOXP2 is essential for speech and communication in people. Svante Paabo and his team at the Max Planck Institute in Leipzig, Germany, discovered this by analyzing the amino acid pattern of Human FOXP2 with that of other creatures (Enard et al. 2002). The discovery that a British family with a severe hereditary speech impediment had an adapted version of the FOXP2 gene supports this hypothesis (Lai et al. 2001, 2003). The human FOXP2 gene differs from those found in chimps and other monkeys (Enard et al. 2002). The mutant human FOXP2 gene, like the chimp's FOXP2, was discovered in a British family with speech problems.

The examination of the Neanderthal genome sequencing revealed a significant fact: they have a similar nucleotide sequence in the FOXP2 gene as people. This discovery implies that Neanderthals could communicate. The FOXP2 gene, which is prevalent in humans and Neanderthals, is thought to have evolved 400,000 years ago in a shared ancestor. Current studies demonstrate that transgenic mice with even a personal version of the FOXP2 gene communicate differently, indicating that the FOXP2 gene is involved in speech regulation. Studies on such transgenic mice also point to the significance of additional genes in human voice regulation in addition to the FOXP2 gene.

The absence of a hereditary kind of standup walking potential has been discovered in a Turkish community. Nevertheless, researchers have discovered an uncommon gene mutation that causes individuals to lose their right to move straight. This gene regulates a protein involved in the formation of the cerebellum in the mind. This gene mutation might decrease speech and mental skills in addition to monitoring vertical gait. Other genes may be involved in influencing the capacity to walk bipedally upright in humans, according to a study of identical individuals in Iraq and Brazil.

Humans also have significantly bigger brains. The ASPM gene, which influences human brain size, has been identified. Moreover, the genes HAR1 and HAR2, which are crucial in deciding human distinctiveness, have recently been identified. These genes are known as HAR (human accelerated region) since they are situated in an area with significant mutation activity; for instance, the HAR-1 DNA sequence exhibits just one mutation variation in chimps and mice, but 32 mutations in humans and chimps for the same length of DNA. At minimum two of these genes, HAR-1 and HAR-2, have been identified in humans. The transcription factor HAR-1 is essential for the onset of the human cortex's six-layer organization. In humans, HAR-2 is involved in the regulation of the hands. Aside from such variations, humans are the only species to have various variants of the amylase gene, as well as a type of the lactose gene which permits an adult person to digest lactose, or "milk sugar." These genes provide people with nutritional benefits. Human and chimp genomes have 3 billion base units in common and are 99 percent similar. There is a 1.23 percent discrepancy, which equals 35 million DNA base pairs. Moreover, the human genome has around 5 million losses as well as penetrations of DNA sequences. As a result, the genomes of humans and chimps vary by around 4%. Approximately 580 genes out of the 25,000 have experienced quick and positive changes as the FOXP2 gene has. HARs are also among such genes. All of such genes are found in DNA sequences that serve no use. In addition, the chimp genome lacks several such genes. Three essential genes that govern inflammation, for instance, are absent from the chimp genome, explaining why humans and chimps have different immunological and inflammatory responses. Similarly, the human genome is lacking numerous genes. This contains the caspase-12 gene, which protects chimps from Alzheimer's disease yet induces Alzheimer's disease in humans when it is absent. Another significant finding of the compared research of human and chimpanzee genomes is that evolving an organism does not need a large number of genomic alterations; rather, a small number of mutations may create a new creature (Pollard 2009).

2.2.4 Structural Genomics, Functional Genomics, and Comparative Genomics

Genome studies have many purposes. These tasks involve identifying their nucleotide sequences, defining their functionality, and comparing them to determine their evolutionary pattern. Structure genomics, functional genomics, and comparative genomics are the three fields of genomics. As a result, structural genomics entails determining an organism's whole nucleotide sequence. Additional than 500 species' whole nucleotide sequences have been obtained, and roughly 3000 more organisms' sequences are being studied.

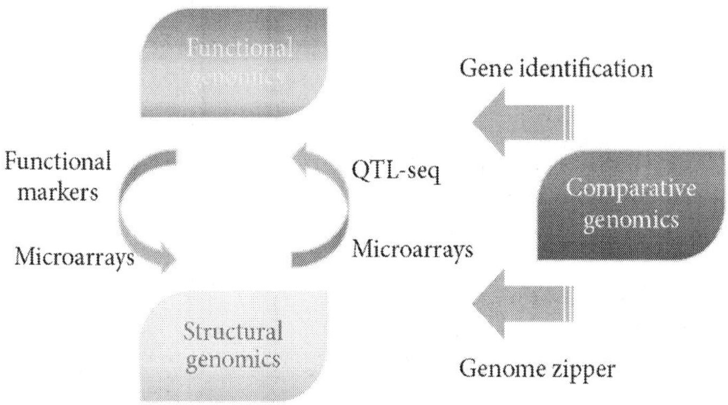

Figure 2.7. Functional, structural, and comparative genomics techniques are all interconnected.

Source: Akpinar, Ani & Lucas, Stuart & Budak, Hikmet. (2013). Genomics Approaches for Crop Improvement against Abiotic Stress. TheScientificWorldJournal. 2013. 361921. 10.1155/2013/361921

The study of the result of various DNA sequences in any species is known as functional genomics. It tries to analyze the function of around 25,000 genes in humans, as well as noncoding and repetitive sequences. It also tries to figure out what function SNPs play in the genome by studying the differences in the reaction of various copies of the similar DNA sequences or genes to diverse pharmacological, chemical, or disease sensitivities. Pharmacogenomics is the study of how drugs react differently in different people.

The study of genomes from various species is known as comparative genomics. Typically, the sequence homology of their genomes is calculated. This involves comparing a gene's introns and exons, the genome's internal transcribed noncoding regions, and gene placement on the chromosomal of the animals being studied. Its major goal is to figure out how various creatures are related evolutionarily. A software tool named Basic Local Alignment Search Tool determines the sequence similarity (BLAST). This software tool compares nucleotide or protein patterns to a data set and calculates the statistical validity of similarities to discover the areas of local homology. By finding both operational and historical links across sequences, BLAST may also be utilized to allocate a specific sequence to a family of genes.

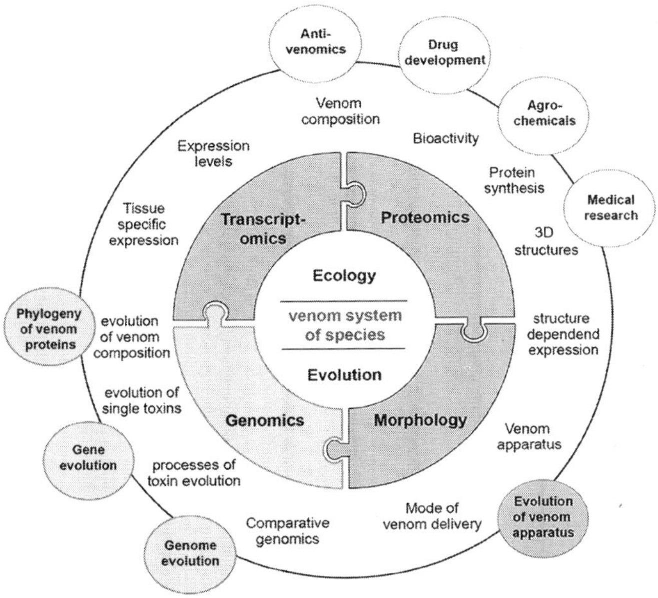

Figure 2.8. Comparative genomics may aid in the prediction of unidentified physiological and metabolic genes' activity-here shown for venomics.

Source: Drukewitz SH and von Reumont BM (2019) The Significance of Comparative Genomics in Modern Evolutionary Venomics. Front. Ecol. Evol. 7:163. doi: 10.3389/fevo.2019.00163

2.2.5 Technical Advantages of Genomics

The growth of advanced methods enabled the satisfactory conclusion of genome sequencing and numerous other genome-related studies. Such techniques comprise mass spectrometry, DNA/RNA microarrays, protein chips, and mass sampling. Typically, such technologies create an enormous volume of data in a brief quantity of time. Many robotics- and computer-based high-throughput technologies have assisted the analysis of DNA at a massive scale. 454 Life Sciences created this one approach (Branford, CT). This technology utilizes picolitre reactors equipped with robotic fluidic as well as optical devices for combining sequenced reagents and detecting the result of the sequencing process (Margulies et al. 2005). This method analyzes around 20 million bases in nearly 4 hours. This approach produces a unique collection of DNA fragments from the complete genome, separates single-stranded DNA molecules on a bead, multiplies and identifies them using PCR, and then analyzes the DNA fragments. This approach does not need subcloning of DNA fragments in bacteria or single clone manipulation. This approach employs picoliter-volume pyrosequencing on a solid substrate. Pyrosequencing involves synthesis-based analyzes. In this technique, a template DNA segment is replicated, and each period a nucleotide is introduced to the side chain, a group of enzymatic reactions provokes chemoluminescence or a light massage that is seized by a charge-coupled device (CCD) camera and evaluated by a computer algorithm to create the nucleotide sequence. This technique provides a significant improvement over all techniques that use Sanger's approach (Sanger et al. 1977). VisiGen Biotechnologies, Inc. (Houston, TX) is creating a DNA sequencing device capable of analyzing 1 million characters per second. By 2010, this equipment will be accessible. This device, like some of the other throughput technologies, is premised on the idea of sorting by synthesized. This will analyze a 3.3 billion-base genetic code in one hour. This tool will be very beneficial for clinical applications and customized treatment.

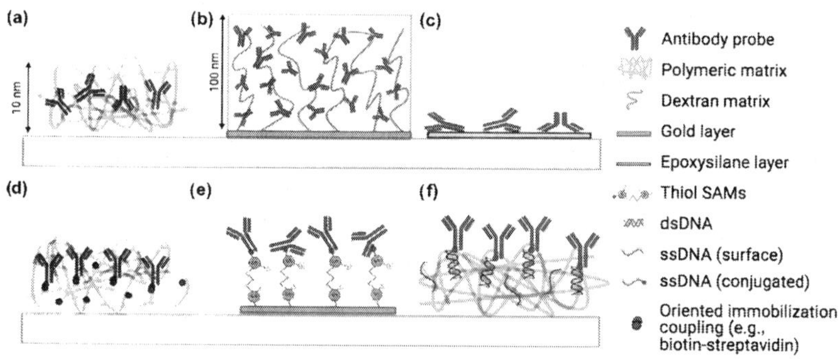

Figure 2.9. Functional Genome Illustration.

Source: https://www.mdpi.com/2073-4360/13/7/1026

DNA microarray is the other high-throughput technique for determining the transcription of a group of genes in diverse circumstances in a single experiment. In this procedure, DNA sequences covering all or the majority of an individual's genetics are first positioned at a specified location on a glass chip, which is then hybridized with cDNA or mRNA from a cell. By attaching to two separate fluorophores, like rhodamine (red) and fluorescein (green), the cDNA derived from mRNA from two distinct cell lines or even a cell line cultured over two distinct growth circumstances is color-tagged (green). The combination of colored cDNA with Target DNA on a glass chip allows for the processing of various hues. cDNA from mRNA generated from yeast maintained in anaerobic and aerobic circumstances, for instance, will only display a red color for gene products under oxygenated settings but would display a green color for genes shown under anaerobic environments. The genes produced in either aerobic or anaerobic situations will appear in yellow. Similarly, a normal cell might create red colors, as well as a cancerous cell, as well as a cancerous cell, might generate green colors for the genetics solely produced by such two cell lines. Nevertheless, the genes produced by the cell lines will appear in yellow. This technique was created by Schena and colleagues at Stanford University in 1995 (Schena et al., 1995), and it has undergone significant development since then. Southern hybridization relies on DNA–DNA hybridization to create a DNA array for analyzing gene expression (Southern 1975). Using the existing known methods, a chip holding the whole genome of an organism, such as 6000 yeast genes, may be manufactured with relative ease. In this kind of DNA matrix study, the color intensity may be utilized to assess the gene expression.

A protein matrix has been designed to determine a cell's protein expression patterns. Based on protein-protein interactions, like antigen/antibody connections employed in Western hybridization. A significant quantity of antibodies or other ligands are placed on particular spots on a glass chip, and their association with rhodamine- or fluorescently protein lysates representing cell lines or just a cell line cultured in two distinct circumstances is then evaluated. The outcomes are represented by red, green, and yellow dots. Likewise, the brightness of color may be utilized to denote the protein levels of expression. Protein chips are very useful in proteomics for analyzing protein-protein relationships, protein changes, and even identifying enzyme substrates. For several of these investigations, the protein chip is generated by fusing ORF transcripts of a cell with glutathione. This approach is very useful for drug development and the identification of proteins associated with disease management.

Mass spectrometry is the final in a succession of high-throughput techniques that have accelerated the detection of proteins by identifying the structure of peptides and matching them to the proteins expressed by the target DNA in the genome central database. The whole procedure is completed in a matter of minutes utilizing nanoscale protein quantity. A particular protein molecule or the entire set of proteins out of a cell is absorbed enzymatically, and afterward the volume of the arising peptide is ascertained by mass proportion in a mass spectrometer and paired with the protein pattern in a protein database to confirm the origin of the protein.

REFERENCES

1. Asgari, E., &Mofrad, M. R. (2015). Continuous distributed representation of biological sequences for deep proteomics and genomics. *PloS one*, *10*(11), e0141287.
2. Baak, J. P. A., Path, F. R. C., Hermsen, M. A. J. A., Meijer, G., Schmidt, J., & Janssen, E. A. M. (2003). Genomics and proteomics in cancer. *European journal of cancer*, *39*(9), 1199-1215.
3. Beadle, G. W. and E. L. Tatum, 1941. Genetic control of biochemical reactions in Neurospora. Proc. Natl. Acad. Sci. U.S.A. 27, 499–506.
4. Bendixen, E. (2005). The use of proteomics in meat science. *Meat science*, *71*(1), 138-149.
5. Benesch, J. L., Ruotolo, B. T., Simmons, D. A., & Robinson, C. V. (2007). Protein complexes in the gas phase: technology for structural genomics and proteomics. *Chemical reviews*, *107*(8), 3544-3567.
6. Botstein, D., R. L. White, M. Skolnick, and R. W. Davis. 1980. Construction of genetic linkage in man using restriction fragment length polymorphism. Am. J. Hum. Genet. 32, 314.
7. Britten, R. J., and D. E. Kohne. 1968. Repeated sequences in DNA. Hundreds of thousands of copies of DNA sequences have been incorporated into the genomes of higher organisms. Science 161, 529–540.
8. Cahill, D. J., Nordhoff, E., O'Brien, J., Klose, J., Eickhoff, H., &Lehrach, H. (2001). Bridging genomics and proteomics. *Proteomics: from protein sequence to function. BIOS Scientific Publishers, Oxford*, 1-22.
9. Carpi, A., Mechanick, J. I., Saussez, S., &Nicolini, A. (2010). Thyroid tumor marker genomics and proteomics: diagnostic and clinical implications. *Journal of cellular physiology*, *224*(3), 612-619.
10. Collins, F., and D. Galas. 1993. A new five-year plan for the U.S. human genome project. Science 262, 43–46.
11. Cox, J., & Mann, M. (2007). Is proteomics the new genomics?. *Cell*, *130*(3), 395-398.
12. de Groot, A., Dulermo, R., Ortet, P., Blanchard, L., Guérin, P., Fernandez, B., ... &Armengaud, J. (2009). Alliance of proteomics and genomics to unravel the specificities of Sahara bacterium Deinococcusdeserti. *PLoS genetics*, *5*(3), e1000434.

13. Di Michele, M., Thys, C., Waelkens, E., Overbergh, L., D'Hertog, W., Mathieu, C., ... &Freson, K. (2011). An integrated proteomics and genomics analysis to unravel a heterogeneous platelet secretion defect. *Journal of proteomics*, *74*(6), 902-913.
14. Elango, N., B. G. Hunt, M. A. D. Goodisman, and S. V. Yip. 2009. DNA methylation is widespread and associated with differential gene expression in castes of the honeybee, *Apis mellifera*. *Proc. Natl. Acad. Sci. U.S.A.* (in press).
15. Enard, W., M. Przeworski, S. E. Fisher, C. S. L. Lai, V. Wiebe, T. Kitano, A. P. Monaco, and S. Paabo. 2002. Molecular evolution of FOXP2, a gene involved in speech and language. *Nature* 418, 869–872.
16. Fabian, T. K., Fejerdy, P., &Csermely, P. (2008). Salivary genomics, transcriptomics and proteomics: the emerging concept of the oral ecosystem and their use in the early diagnosis of cancer and other diseases. *Current genomics*, *9*(1), 11-21.
17. Fermin, D., Allen, B. B., Blackwell, T. W., Menon, R., Adamski, M., Xu, Y., ... &Omenn, G. S. (2006). Novel gene and gene model detection using a whole genome open reading frame analysis in proteomics. *Genome biology*, *7*(4), 1-13.
18. Fry, B. G. (2005). From genome to "venome": molecular origin and evolution of the snake venom proteome inferred from phylogenetic analysis of toxin sequences and related body proteins. *Genome Research*, *15*(3), 403-420.
19. Gstaiger, M., &Aebersold, R. (2009). Applying mass spectrometry-based proteomics to genetics, genomics and network biology. *Nature Reviews Genetics*, *10*(9), 617-627.
20. Günther, O. P., Shin, H., Ng, R. T., McMaster, W. R., McManus, B. M., Keown, P. A., ... &Lê Cao, K. A. (2014). Novel multivariate methods for integration of genomics and proteomics data: applications in a kidney transplant rejection study. *Omics: a journal of integrative biology*, *18*(11), 682-695.
21. Gupta, A. K., Kaur, K., Rajput, A., Dhanda, S. K., Sehgal, M., Khan, M., ... & Kumar, M. (2016). ZikaVR: an integrated Zika virus resource for genomics, proteomics, phylogenetic and therapeutic analysis. *Scientific reports*, *6*(1), 1-16.
22. Hall, N., Karras, M., Raine, J. D., Carlton, J. M., Kooij, T. W., Berriman, M., ... &Sinden, R. E. (2005). A comprehensive survey of

the Plasmodium life cycle by genomic, transcriptomic, and proteomic analyses. *Science*, *307*(5706), 82-86.

23. Hanash, S. M. (2001). Global profiling of gene expression in cancer using genomics and proteomics. *Current Opinion in Molecular Therapeutics*, *3*(6), 538-545.
24. Hogeweg, P. 1978. Simulating the growth of cellular forms. *Simulation* 31, 90–96.
25. Hogeweg, P. and B. Hesper. 1978. Interactive instruction on population interactions. *ComputBiol* Med 8, 319–327.
26. Ibrahim, S. M., & Gold, R. (2005). Genomics, proteomics, metabolomics: what is in a word for multiple sclerosis? *Current opinion in neurology*, *18*(3), 231-235.
27. inhibitors. Proc. Nat. Acad. Sci. U.S.A. 74, 5463–5467.
28. Ju, C., Feng, Z., Brindley, P. J., McManus, D. P., Han, Z., Peng, J. X., & Hu, W. (2010). Our wormy world: genomics, proteomics and transcriptomics in East and Southeast Asia. *Advances in parasitology*, *73*, 327-371.
29. Kan, Y. W. and A. M. Dozy. 1978. Polymorphism of DNA sequence adjacent to human beta-globin gene: Relationship to sickle mutation. Proc. Natl. Acad. Sci. U.S.A. 75, 5637.
30. Kenyon, G. L., DeMarini, D. M., Fuchs, E., Galas, D. J., Kirsch, J. F., Leyh, T. S., ... &Sheahan, L. C. (2002). Defining the mandate of proteomics in the post-genomics era: workshop report. *Molecular & Cellular Proteomics*, *1*(10), 763-780.
31. Kovac, J. R., Pastuszak, A. W., & Lamb, D. J. (2013). The use of genomics, proteomics, and metabolomics in identifying biomarkers of male infertility. *Fertility and sterility*, *99*(4), 998-1007.
32. Lai C. S., D. Gerrelli, A. P. Monaco, S. E. Fisher, A. J. Copp. 2003. FOXP2 expression during brain development coincides with adult sites of pathology in a severe speech and language disorder. *Brain* 126, 2455–2462.
33. Lai, C. S., S. E. Fisher, J. A. Hurst, F. Vargha-Khadem, A. P. Monaco. 2001. A forkhead-domain gene is mutated in a severe speech and language disorder. *Nature* 413, 519–523.
34. Lan, N., Montelione, G. T., & Gerstein, M. (2003). Ontologies for proteomics: towards a systematic definition of structure and

function that scales to the genome level. *Current opinion in chemical biology*, *7*(1), 44-54.

35. Latonen, L., Afyounian, E., Jylhä, A., Nättinen, J., Aapola, U., Annala, M., ... &Visakorpi, T. (2018). Integrative proteomics in prostate cancer uncovers robustness against genomic and transcriptomic aberrations during disease progression. *Nature communications*, *9*(1), 1-13.
36. Li, H. D., Menon, R., Omenn, G. S., & Guan, Y. (2014). Revisiting the identification of canonical splice isoforms through integration of functional genomics and proteomics evidence. *Proteomics*, *14*(23-24), 2709-2718.
37. Low, T. Y., van Heesch, S., van den Toorn, H., Giansanti, P., Cristobal, A., Toonen, P., ... & Guryev, V. (2013). Quantitative and qualitative proteome characteristics extracted from in-depth integrated genomics and proteomics analysis. *Cell reports*, *5*(5), 1469-1478.
38. Margulies, M., M. Egholm, W. E. Altman, S. Attiya, J. S. Bader, L. A. Bemben, J. Berka, M. S. Braverman, Y. J. Chen, Z. Chen et al.. 2005. Genome sequencing in microfabricated high-density picolitre reactors. *Nature* 437, 376–380.
39. Martyniuk, C. J., &Denslow, N. D. (2009). Towards functional genomics in fish using quantitative proteomics. *General and comparative endocrinology*, *164*(2-3), 135-141.
40. Maxam, A. M. and W. Gilbert. 1977. A new method for sequencing DNA. *Proc. Natl. Acad. Sci.* U.S.A. 74, 560.
41. McLean, T. I. (2013). "Eco-omics": a review of the application of genomics, transcriptomics, and proteomics for the study of the ecology of harmful algae. *Microbial ecology*, *65*(4), 901-915.
42. Miller, I., Rogel-Gaillard, C., Spina, D., Fontanesi, L., & M de Almeida, A. (2014). The rabbit as an experimental and production animal: from genomics to proteomics. *Current Protein and Peptide Science*, *15*(2), 134-145.
43. Mullis, K. B. and F. A. Faloona. 1987. Specific synthesis of DNA *in vitro* via a polymerase catalyzed chain reaction. Methods Enzymol. 155, 335.
44. Mustafa, A. S. (2005). Mycobacterial gene cloning and expression, comparative genomics, bioinformatics and proteomics in relation to the development of new vaccines and diagnostic reagents. *Medical Principles and Practice*, *14*(Suppl. 1), 27-34.

45. Oldham, P. D. (2009). Global status and trends in intellectual property claims: genomics, proteomics and biotechnology. *Proteomics and Biotechnology (January, 22 2009)*.

46. Özdemir, V., Dove, E. S., Gürsoy, U. K., Şardaş, S., Yıldırım, A., Yılmaz, Ş. G., ... & Srivastava, S. (2017). Personalized medicine beyond genomics: alternative futures in big data—proteomics, environtome and the social proteome. *Journal of neural transmission*, *124*(1), 25-32.

47. Pardanani, A., Wieben, E. D., Spelsberg, T. C., &Tefferi, A. (2002, November). Primer on medical genomics part IV: expression proteomics. In *Mayo Clinic Proceedings* (Vol. 77, No. 11, pp. 1185-1196). Elsevier.

48. Pognan, F. (2004). Genomics, proteomics and metabonomics in toxicology: Hopefully not 'fashionomics'. *Pharmacogenomics*, *5*(7), 879-893.

49. Pollard, K. S. 2009. What makes us human? Sci. Am. 300, 44–49.

50. Rao, K. D., &Swamy, M. N. S. (2008). Analysis of genomics and proteomics using DSP techniques. *IEEE Transactions on Circuits and Systems I: Regular Papers*, *55*(1), 370-378.

51. Rodrigues, P. M., Silva, T. S., Dias, J., &Jessen, F. (2012). Proteomics in aquaculture: applications and trends. *Journal of proteomics*, *75*(14), 4325-4345.

52. Sali, A., Glaeser, R., Earnest, T., &Baumeister, W. (2003). From words to literature in structural proteomics. *Nature*, *422*(6928), 216-225.

53. Sanger. F., S. Nicklen, and A. R. Coulson. 1977. DNA sequencing with chain terminating

54. Schena, M., D. Shalon, R. W. Davis, and P. O. Brown. 1995. Quantitative monitoring of gene expression patterns with a complementary DNA microarray. Science 270, 567–570.

55. Skylas, D. J., Van Dyk, D., & Wrigley, C. W. (2005). Proteomics of wheat grain. *Journal of cereal science*, *41*(2), 165-179.

56. Smith, L. M., Fung, S., Hunkapiller, M. W., Hunkapiller, T. J., and Hood, L. E. 1985. The synthesis of oligonucleotides containing an aliphatic amino group at the 5_ terminus: Synthesis of fluorescent DNA primers for use in DNA sequence analysis. *Nucleic Acids Res.* 13, 2399–2412.

57. Solé, R. V., & Pastor-Satorras, R. (2002). Complex networks in genomics and proteomics. *Handbook of Graphs and Networks*, 145-167.
58. Song, X., & Lin, Q. (2017). Genomics, transcriptomics and proteomics to elucidate the pathogenesis of rheumatoid arthritis. *Rheumatology international*, *37*(8), 1257-1265.
59. Southern, E. M. 1975. Detection of specific sequences of DNA fragments separated by gel electrophoresis. J Mol. Biol. 98, 503.
60. Sullivan, M. B., Krastins, B., Hughes, J. L., Kelly, L., Chase, M., Sarracino, D., & Chisholm, S. W. (2009). The genome and structural proteome of an ocean siphovirus: a new window into the cyanobacterial 'mobilome'. *Environmental Microbiology*, *11*(11), 2935-2951.
61. Vaidyanathan, P. P., & Yoon, B. J. (2004). The role of signal-processing concepts in genomics and proteomics. *Journal of the Franklin Institute*, *341*(1-2), 111-135.
62. Wright, J. C., Beynon, R. J., & Hubbard, S. J. (2010). Cross species proteomics. In *Proteome Bioinformatics* (pp. 123-135). Humana Press.
63. Yates III, J. R. (2000). Mass spectrometry: from genomics to proteomics. *Trends in genetics*, *16*(1), 5-8.
64. Zivanovic, Y., Armengaud, J., Lagorce, A., Leplat, C., Guérin, P., Dutertre, M., ... &Confalonieri, F. (2009). Genome analysis and genome-wide proteomics of Thermococcusgammatolerans, the most radioresistant organism known amongst the Archaea. *Genome biology*, *10*(6), 1-23.
65. Zivy, M., & de Vienne, D. (2000). Proteomics: a link between genomics, genetics and physiology. *Plant Molecular Biology*, *44*(5), 575-580.

CHAPTER 3

METHODOLOGY FOR SEPARATION AND IDENTIFICATION OF PROTEINS AND THEIR INTERACTIONS

CONTENTS

3.1 INTRODUCTION

In this lesson, we will cover techniques for (a) the separation of proteins, (b) the discovery of the basic protein molecule by identification of their amino acid patterns, (c) the 3D structure, of proteins, and (d) the number of proteins at the proteome scale.

A cell has a multitude of proteins. The constituent proteins are isolated and characterized to know the function of distinct proteins plus their functional and structural interactions. Proteins are extracted by some techniques as they have a wide range of characteristics (Abdelhamid & Wu, 2015; Field et al., 2020).

3.2 SEPARATION OF PROTEIN VIA THE MULTIDIMENSIONAL APPROACH

As discussed in Chapter 1, a multidimensional strategy is better for separating them as compared to the single one. A multifaceted method like this must handle issues like resolution, throughput, automation, and adaptation to mass spectrometry analysis. Electrophoresis, which includes two-dimensional (2D) gel electrophoresis plus capillary electrophoresis, is among the most essential methods for separating them. After these technologies have separated them, mass spectrometry is used to identify them. This section elaborates on such techniques (Salzer et al., 2008; Büyükköroğlu et al., 2018).

3.1.1 Electrophoresis

Molecules may be sorted depending on their electrostatic forces in an electric field. This technique is known as electrophoresis, and it was discovered by Swedish scientist Arne Teselius, who was given the 1948 Nobel Prize for this achievement. A solid substrate, like paper or gel, is coated with molecules, which are subsequently exposed to an electrical field. The motion of the molecules in an electric field is determined by their electrical charges. Typically, this separates proteins using a gel matrix, as shown below.

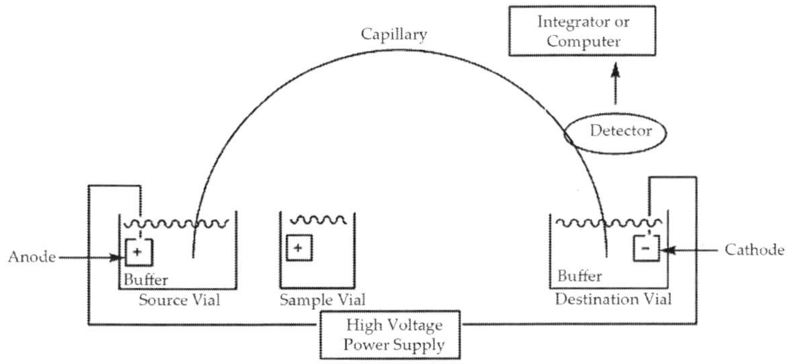

Figure 3.1. Basic Electrophoresis Principle.

Source: By Apblum - http://en.wikipedia.org/wiki/Capillary_electrophoresis, CC BY-SA 3.0, https://commons.wikimedia.org/w/index.php?curid=35013009

2D Gel Electrophoresis is a technology invented separately by O'Farrell (1975) and Klose (1975) for separating proteins (Figure 3.1). Proteins in the *Escherichia coli* cell isolate were isolated by gel electrophoresis conducted on two surfaces that were perpendicular to one another. Initially, the proteins were extracted according to their charges, then they were sorted according to their molecular weights in a vertical plane to the first layer. Following two electrophoretic cycles, the solution was mixed to see the protein bands. Thus, over 1100 protein bands were recovered from the whole-cell extract of *E. coli*. The 2D gel technique has also been referred to as the Iso-Dalt method, since it separates proteins depending on their differences in electrical charge and mass, as indicated by isoelectrofocusing (IEF) [isoelectric point (pI)] and Dalton, the unit of mass (Arima & Iwata, 2007; Jorrin-Novo, 2014).

Electrophoresis is a typical technique for isolating molecules, particularly proteins, according to their charge-to-mass ratio (e/m) and the intensity of the electric field. When a protein composition is exposed to an electric field, the protein molecules migrate into distinct zones depending on their carrier concentration, which is defined by the charge to mass (e/m) proportion of protein complexes (Kay et al., 2000; Turriziani et al., 2016). Typically, it separates proteins on a solid substrate, like polyacrylamide gel. On a solid matrix, isolated protein molecules are kept to their corresponding zones with little diffusion and heat production.

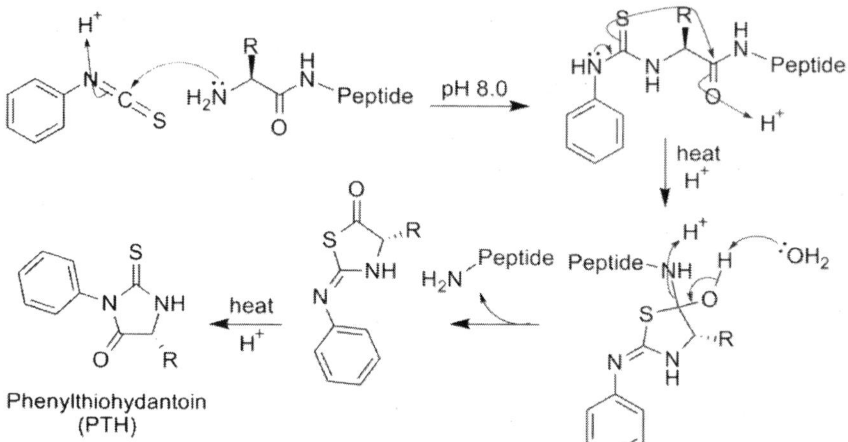

Figure 3.2. Depicts the stages of Edman's degradation. Without hydrolyzing the rest of the peptide, the tagged aminoterminal residue (PTH - alanine in round one) is liberated. Repeating the cycle reveals the entire pattern of the peptide.

Source: https://chem.libretexts.org/Courses/University_of_Arkansas_Little_Rock/CHEM_4320_5320%3A_Biochemistry_1/02%3A__Protein_Structure/2.2%3A_Protein_Sequencing

Heat production presents substantial complications in separating proteins in a liquid matrix. Under particular circumstances employing a small tube containing a combination of proteins in liquid, as in capillary electrophoresis, such challenges of separating proteins are significantly reduced. Polyacrylamide gel electrophoresis (PAGE) is often used to separate proteins since polyacrylamide gel produces a sieving impact throughout separating proteins due to the pore diameter of the gel. In addition, the diameter of the holes may be altered by adjusting the concentration of monomers in the gel, like acrylamide, the gelling agent, and bisacrylamide, the crosslinking agent in the gel. With a higher quantity of acylamide, the hole diameter of the gel decreases. PAGE gels comprising 15 percent acrylamide and 5 percent bisacrylamide are often used to separate proteins. Based on the scale of the proteins, gels having 3 to 30 percent polyacrylamide may be used for protein extraction. For separating proteins with a molecular weight larger than 1 million Daltons, a 3 percent gel is utilized. For the separating proteins with a molecular mass of just under 1000 Daltons, a 30 percent gel is utilized. A 3 percent polyacrylamide gel is fragile and complicated to manipulate; these issues are typically eliminated by the addition of agarose all through gel formation. Agarose stabilizes the gel with not impairing the movement of the gel's proteins (Meyer & Peters, 2003; Mahmoudi et al., 2011).

The 2D gel or Iso-Dalt is the most popular mechanism in proteomics due to its relative ease of using it, robotization, good reproducibility, resolving power of proteins, and mass spectrometry validity. In addition, protein bands separated by Iso-Dalt are amenable to Edman degradation and amino acid structure assessment (Chouchani et al., 2011). As previously stated, the 2D gel technique separates proteins by resolution depending on the mass and charge of the proteins in two distinct aspects. Separation depending on this double property of proteins offers superior resolution by ignoring the existence of multiple proteins per band and/or cross-contamination of a protein band through other protein parts and by allowing the visualization of proteins that are present in minute quantities in the cell or body fluid. To obtain a high resolving power of newly synthesized proteins in proteomics, it is necessary, to begin with, a test tube containing all of the proteins introduced in a cell or body fluid. Care must be taken to involve every protein membrane that exists in the cell or body fluid or other hydrophobic proteins, rare protein parts, and proteins with large differences in charge material, like those, for example with a pH below 3 or over 10 (Brohee& Van Helden, 2006; Cedervall et al., 2007).

The 2D gel technique for separating proteins has been discussed in detail by Gorg and others (1988), and Gorg himself is credited with its development (2000). The following phases include the most essential features of this technique:

- Preparing the specimen, solubility, and applying to the gel
- Protein segregation on an implanted pH gradient (IPG) gel strip using IEF
- On a sodium dodecyl sulfate (SDS)-PAGE, the proteins in the IPG strip were separated.
- After SDS-PAGE, observation of divided protein bands and assessment of the structure of the divided proteins on the gel

3.1.2 Liquid Chromatography

Chromatography is the isolation of proteins or peptides by putting a mixture of proteins or peptides in a suitable solvent more than a solid matrix. Several types of chromatography exist dependent on the type of the matrix, including liquid or column chromatography, paper chromatography, thin gel chromatography, and gas chromatography. The type of chromatography most applicable to proteome analysis as well as appropriate for mass spectrometric analysis is liquid chromatography, also recognized as column chromatography since the matrix comprised of beads packed as a column in a glass tube whereby the protein or peptide solution is passed, resolved depending on their size, charge, or affinity to a ligand, and segments of divided proteins or peptides are gathered (Haque, 1998; Haque, 1998).

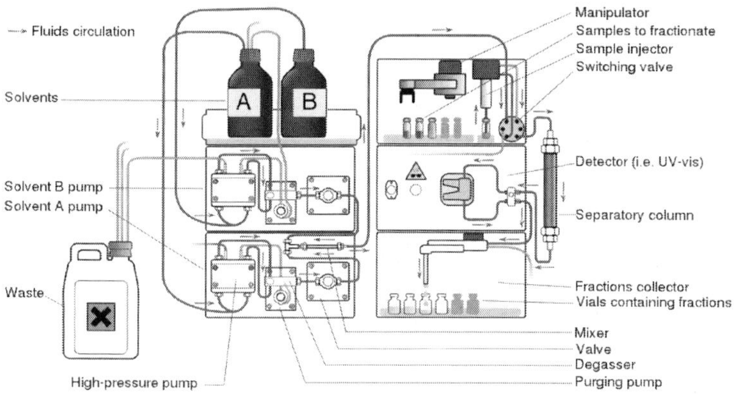

Figure 3.3. Typical route stream in liquid chromatography.

Source: By This W3C-unspecified vector image was created with Inkscape. - Own work, CC BY 3.0, https://commons.wikimedia.org/w/index.php?curid=10988849

3.1.2.1 Gel Filtration

Size-exclusion chromatography, often known as gel filtration, is a kind of liquid chromatography that separates proteins based on their size. Proteins are isolated using this approach depending on their molecular size. A protein solution is handed down a column of neutral materials having a set pore size, like agarose or Sephadex, and afterward eluted with a buffer. Oversized proteins that are unable to penetrate the hole of the beads snake thru the gap across beads and are ejected from the column throughout the elution phase. As a result, bigger proteins emerge first, well ahead of smaller proteins. Smaller proteins, on the other hand, penetrate the bead hole and require significantly longer to exit the holes upon elution, therefore they are eluted considerably later than bigger proteins. Following elution, the distinct proteins are isolated and gathered as separate fractions (Zölls et al., 2012; Sorci et al., 2013).

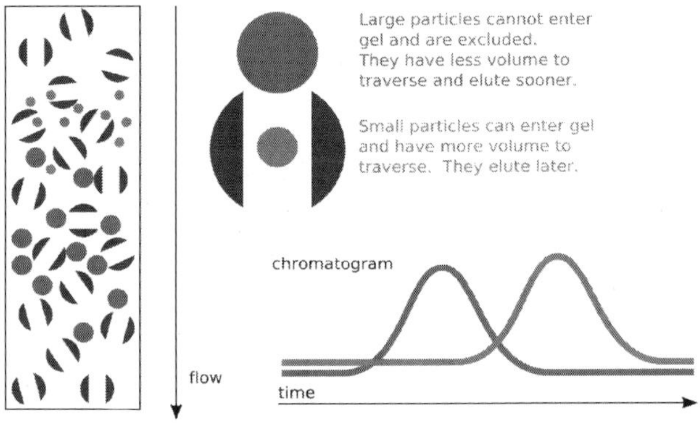

Figure 3.4. Gel filtering is used to separate molecules of various sizes.

Source; https://commons.wikimedia.org/wiki/File:SizeExChrom.png

3.1.2.2 Affinity Chromatography.

Affinity chromatography is liquid chromatography that isolates proteins depending on their affinity for a ligand linked to the matrix. In this procedure, proteins without affinity for the matrix-bound ligand stay unattached and are easily eluted from the column. In comparison, a specific protein in the combination may attach to a ligand, delaying its clearance from the column; this protein gets deleted from the column later due to a variation in elution circumstances. This chromatographic technique is utilized for the separation of proteins that stay bound to the ligand connected to the column from the majority of a protein. Some proteins including histidine oligopeptides that are eliminated on a nickel column serve as an excellent illustration of protein purification via affinity chromatography (Reinders et al., 2006; Urey et al., 2016). Additionally, glutathione-S-transferase-containing proteins may be kept and isolated specifically from a column having glutathione-coated beads. An affinity tag may be developed specifically to separate or clear a certain protein. For instance, proteins required for DNA interactions, like Replication of DNA, maintenance, and crossover, may be isolated using a matrix containing DNA as an affinity tag. Alternately, a protein may be purified by passing it across a matrix having an antibody with an affinity tag linked to it.

3.1.2.3 Ion Exchange Chromatography

Ion exchange chromatography is a kind of liquid chromatography that differentiates proteins depending on electrical charges. Proteins are adsorbed onto a charged pair linked to the cellulose substrate in this procedure. Organic molecules which are anionic or cationic make up the charged pairs. The most popular cation and anion exchangers utilized as a matrix throughout ion-exchange chromatography are carboxymethyls (CM) cellulose and diethyl aminoethyl (DEAE) cellulose (Issaq et al., 2002; Washburn, 2004).

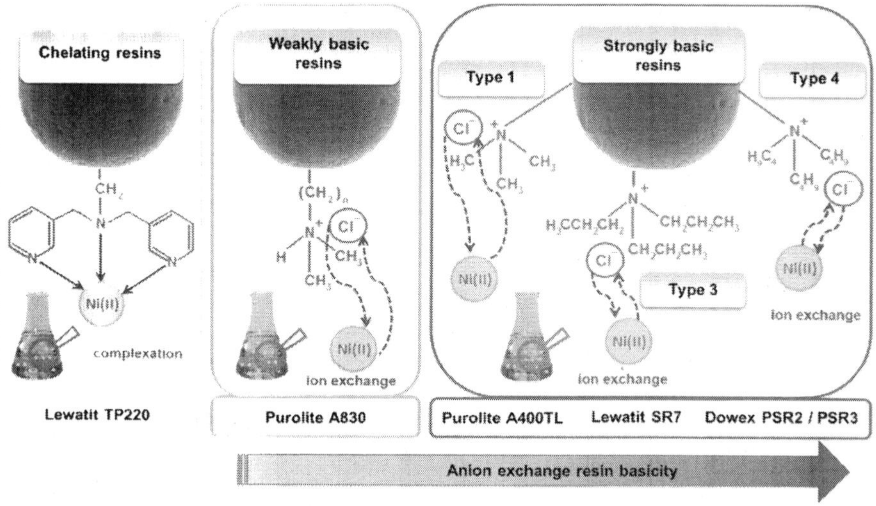

Figure 3.5. Ion-exchange chromatography diagrammatic representation.

Source: Wołowicz A, Wawrzkiewicz M. Screening of Ion Exchange Resins for Hazardous Ni(II) Removal from Aqueous Solutions: Kinetic and Equilibrium Batch Adsorption Method. Processes. 2021; 9(2):285. https://doi.org/10.3390/pr9020285

Along with a variation in the concentration of ions or pH of the elution buffer, the unattached proteins are eliminated first, followed by the bound proteins.

3.3 DETERMINATION OF THE PRIMARY STRUCTURE OF PROTEINS

The main frame of a protein is constituted by the chain of amino acids. This governs how a protein folds and acquires the three-dimensional (3D) shape which regulates a protein's function. Many additional conclusions might be drawn from the amino acid pattern of a protein; for instance, the protein's (pI) might be derived from its amino acid pattern. In addition, the existence of numerous hydrophobic amino acids in the sequencing suggests as it is either a membrane protein or a receptor protein (Link, 2002; Bergström et al., 2006). Additionally, the existence of particular amino acids may signal that protein will develop a beta-sheet structure. Thus, a protein's amino acid pattern and fundamental structure are essential. There are three approaches to identifying a protein's amino acid sequence. Deciphering the nucleotide pattern of a DNA molecule, Edman degradation, and mass spectrometry are examples of such techniques (Issaq et al., 2005; Capriotti et al., 2011).

3.3.1 Proteomics without Spectrometry

3.3.1.1 Determination of Amino acid Sequence from DNA sequence

The amino acid structure of a protein expressed by a gene might be deciphered thanks to our knowledge of gene sequences and our capacity to pattern nucleotides in a DNA part. As long as DNA pattern data was accessible in the GenBank database, it formed common practice to infer the amino acid patterns of various proteins. A considerable percentage of proteins in the protein data bank (PDB) have their main structure determined directly from DNA sequencing (Peng et al., 2008; Acquah et al., 2019).

But it is simple to deduce a protein's amino acid composition from its target DNA, the opposite is not accurate (i.e. decadence of the genetic sequence makes it difficult to deduce the sequence of nucleotides of a gene from the amino acid structure of the protein.). To deduce the pattern of DNA of a gene from the amino acid pattern of a protein, one must depend on the individual's shared use of genetic coding. In molecular biology investigations, it is usual to generate a gene for cloning. In this way, the first insulin gene was created using the amino acid pattern of insulin protein. The polypeptide chain might be deduced directly from the nucleotide sequence in prokaryotes and lesser eukaryotes, like yeast and several filamentous fungi. Due to the huge existence of intron patterns in higher eukaryotes, like mammals, there is no consistent link between nucleotide sequence and amino acid sequencing. To interpret the amino acid composition of a protein as determined by a gene for higher animals, the intron patterns must be disregarded. Moreover, depending on how exons are spliced in higher organisms, multiple proteins may possibly be created. In nature, an organism may not have all of these potentially potential proteins (Wolters et al., 2001; Dixon et al., 2006).

The accessibility of the target DNA of the gene that encodes a protein is sometimes critical in determining the amino acid sequence; for instance, in mass spectrometric analysis, two neighboring glycine residues, including one with a molecular mass of 57 daltons, could show up as asparagines with a molecular mass of 114 daltons. Studying the nucleotide sequence, which shows separate genetic codes for glycine and asparagine, may quickly clarify this dilemma. This kind of nucleotide sequence analysis can determine if a protein's amino acid sequence has two contiguous glycine residues or merely one asparagine residue.

3.3.1.2 Edman Degradation—N-Terminal Amino Acid Sequence Analysis.

The pattern of amino acids was established via Edman degradation until mass spectrometry formed accessible. The process included identifying one amino acid at a period first from the peptide's N-terminus. The initial N-terminal amino acid is treated with phenyl-iso-thiocyanate, and it will be split away by moderate hydrolysis as phenyl hydration, a cyclic compound of the N-terminal amino acid. The truncated peptide is preserved throughout this procedure. The chromatographic profile of the degraded amino acid is used to identify it. The second N-terminal amino acid then is broken, leaving at least one peptide intact but truncated by two amino acids. The chromatographic profile of the degraded amino acid is used to identify it. After each round, the procedure is continued until all of the amino acids have been recognized. This procedure is entirely automated. Nevertheless, it is time taking and tedious. Notwithstanding such inherent challenges, this process, known as Edman degradation and invented by Pehr Edman, was the sole way to identify the amino acid pattern of a protein until mass spectrometry. Here are the stages in the Edman degrading process (Gordon et al., 2013; Cassidy et al., 2021).

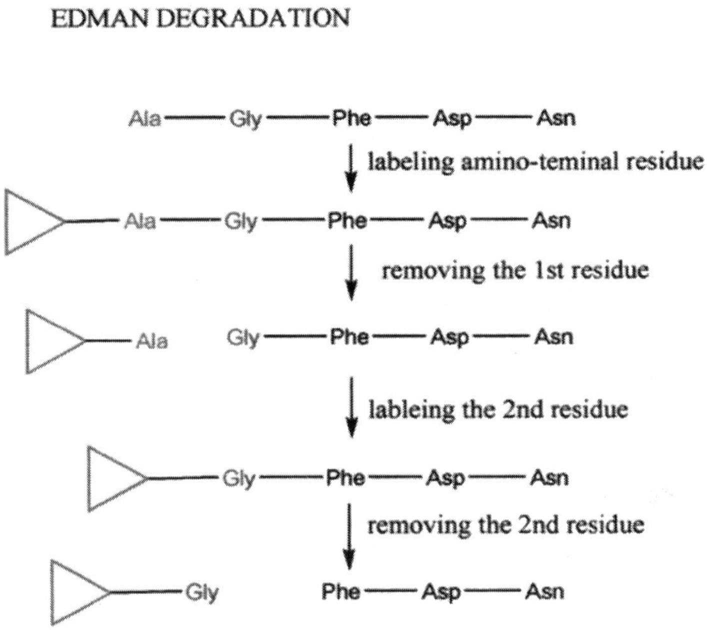

Figure 3.6. Edman degradation, invented by Pehr Edman, may be used to determine the order of amino acids in the protein or peptide.

Source: https://en.wikibooks.org/wiki/Structural_Biochemistry/Proteins/Protein_sequence_determination_techniques

Whenever the 2D gel approach for separating proteins was accessible in the mid-1970s, proteins isolated by 2D gel were frequently put to Edman degradation to identify the amino acid order, since this was the only means to do so at the moment. Nevertheless, there are two issues with this proteomics approach: For starters, several proteins have blocked N-terminal proteins that can't combine with the phenyl isothiocyanate needed for Edman breakdown. Secondly, Edman degradation can only identify the amino acid composition of one protein at a moment, which is in contrast to the goals of proteomics, which are to get data on multiple proteins at once.

3.3.2 Proteomics Based on Mass Spectrometry— Identification of Proteins Based on Their Amino Acid Sequence

After extraction by different techniques, like electrophoresis and/or liquid chromatography, peptides or proteins are recognized by identifying their amino acid pattern. This was formerly accomplished using Edman degradation, which identified one amino acid at a period from the N-terminus of proteins or peptides. Nevertheless, the identification of proteins was transformed by the discovery and deployment of the mass spectrometer, together with breakthroughs in genomics and bioinformatics, who enabled gene and protein data accessible for the attribution of a specific peptide pattern to a protein and its encoding gene.

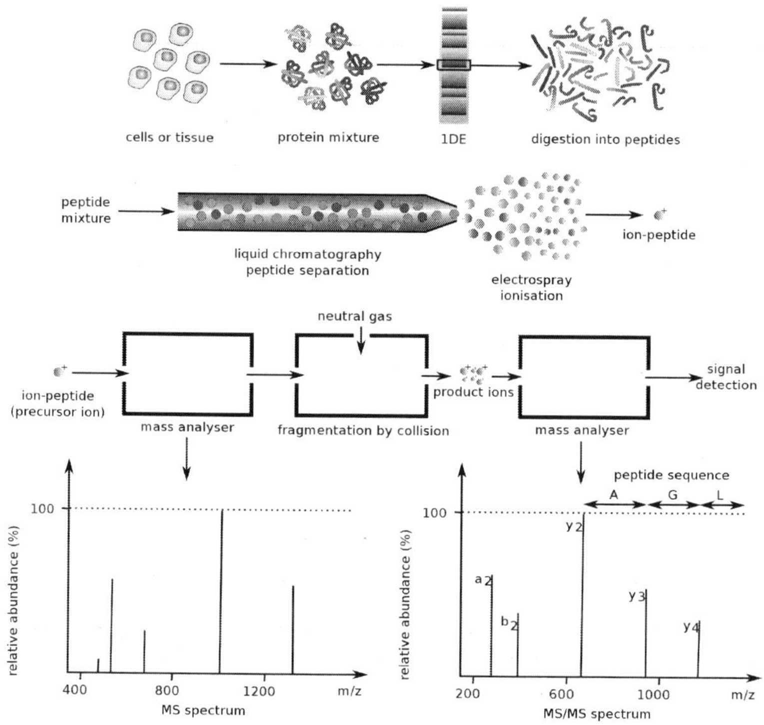

Figure 3.7. Protocol for mass spectrometry.

Source;https://upload.wikimedia.org/wikipedia/commons/thumb/1/1f/Mass_spectrometry_protocol.png/440px-Mass_spectrometry_protocol.png

3.3.2.1 Mass Spectrometry

A mass spectrometer is a useful tool that precisely analyzes the molecular weight of a sample and assists in determining its chemical makeup. That device splits molecules of a sample based on their mass/charge (m/z) relation, providing data on every ion's molecular mass. Their position is formed using this data. Every amino acid has its molecular weight, which provides every peptide its unique molecular weight. The pattern of amino acids in a peptide or protein is deciphered using the molecular mass of the peptide.

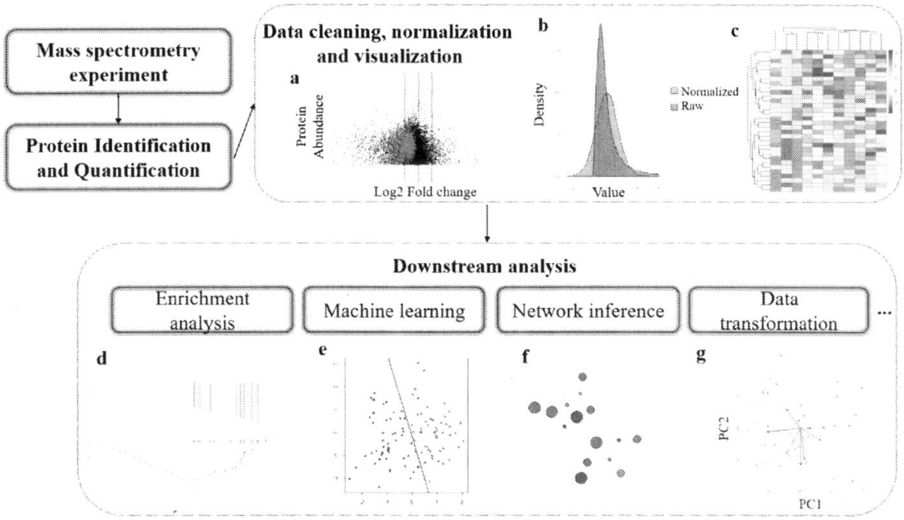

Figure 3.8. Mass Spectrometry Theory.

Source; Chen C, Hou J, Tanner JJ, Cheng J. Bioinformatics Methods for Mass Spectrometry-Based Proteomics Data Analysis. International Journal of Molecular Sciences. 2020; 21(8):2873. https://doi.org/10.3390/ijms21082873

Just a little variation in molecular weight may indicate the substitution of one amino acid for something else or a posttranslational alteration such as phosphorylation, acetylation, or even other structural alteration in a peptide. Table 3.1 shows the molecular mass of various amino acids.

It's worth noting that the molecular mass of amino acids in peptides is lower than those of free amino acids. Since a molecule of water is eliminated throughout most of the peptide synthesis, it is reduced by 18 daltons, which is the molecular weight of water. Proteomics makes extensive utilization of mass spectrometry. This device is being used to define the type of a protein, namely its amino acid pattern and protein-ligand complex creation during physiological settings, such as post-transcriptional alterations, enzyme-substrate bindings, and antigen-antibody or orphan receptors relationships. By measuring the hydrogen/deuterium exchange, this device is also utilized to improve the understanding of protein function.

3.3.2.2 Components of the Instrument.

A spectrometer comprises the following main parts: a connector or gadget for the insertion of a specimen into the device, an ionization gadget, for the splitting of ionized molecules based on their mass to charge (m/z) proportion, a sensor that supervises the split ions and registers them, and a vacuum environment system to enable the free transport of ions inside the spectrometer and analysis software.

Table 3.1. Molecular weights of amino acids in peptides

Amino acids	Symbols—3letter/1letter	Molecular weight
Alanine	ala/A	71
Asparagine	asn/N	114
Aspartate	asp/D	115
Arginine	arg/R	156
Cysteine	cys/C	103
Glutamine	gin/Q	128
Glutamate	glu/E	129
Glycine	gly/G	57
Histidine	his/H	137
Isoleucine	ile/I	113
Leucine	leu/L	113
Lysine	lys/K	128
Methionine	met/M	131
Phenylalanine	phe/F	147
Proline	pro/P	97
Serine	ser/S	87
Threonine	thr/T	101
Tryptophan	trp/W	186
Tyrosine	tyr/Y	163
Valine	val/V	99

The instrument for introducing samples into the system. Based on the sample's composition and technique of ionization, a specimen might well be injected into the spectrometer. A specimen is injected into the spectrometer's ionization resource. Typically, it is injected on a probe, platform, or capillary tube immediately following the HPLC or capillary electrophoresis of peptides and proteins. A specimen has always been entered via a lock mechanism to sustain the machine's high vacuum with no disturbance.

Ionization device. Molecules are polarized in a spectrophotometer since it is simpler to regulate and command the motion of ionized, electrically charged ones than neutral molecules. Typically, ionization is accomplished by protonation (the addition of an H ion) or deprotonation (removal of an H ion). Such two techniques are known as positively and negatively ionization, respectively. Since this NH2 band in protein rapidly takes an H ion, small proteins are often sensitive to positively ionization. Multiple techniques of molecular ionization are put into the device for evaluation. However, two basic types of ionization machines are utilized in proteomics. Among them are the electrospray ionization (ESI) and matrix-aided laser desorption ionization (MALDI) devices. These gadgets are discussed in the sections that follow.

3.3.3 Bottom-Up andTop-Down Mass Spectrometry

"Bottom-up" or peptide-level spectroscopy determines the molecular mass of the peptides formed following proteolysis. Highest or whole protein-level spectrometry, on the other hand, is used to detect intact proteins. Such two approaches are not necessarily incompatible; rather, they are complementary and necessary for obtaining a complete picture of a protein's amino acid composition and various posttranslational modifications. Such two concepts, "bottom-up" and "top-down," are derived from genomics and are similar to those utilized in genetic sequencing for DNA sequencing.

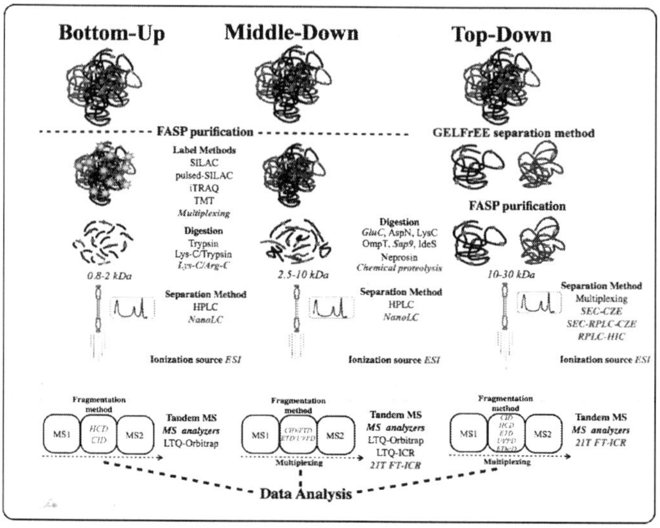

Figure 3.9. Different processes in top-down and bottom-up proteomics are shown.

Source: Gracia KC, Husi H. Computational Approaches in Proteomics. In: Husi H, editor. Computational Biology [Internet]. Brisbane (AU): Codon Publications; 2019 Nov 21. Figure 2, [Bottom–up, middle–down and top–down proteomic...]. Available from: https://www.ncbi.nlm.nih.gov/books/NBK550333/figure/Ch8-f0002/ doi: 10.15586/computationalbiology.2019.ch8

The typical technique for peptide identification is bottom-up spectrometry. The proteins are were broken by bacteria into tiny parts containing 5– 20 amino acids, but then their weight is measured using a spectrometer. To determine the protein, these data are compared to the weights of peptides of a recognized amino acid pattern stored in a database. Nevertheless, this method relies on partial data to present a fuller view of a protein and ensure its recognition. Furthermore, the bottom-up technique is unable to give details on a protein's posttranslational modifications, which is necessary to fully understand its activity, especially its enzymatic involvement in a metabolic circuit. The top-down approach to protein study starts with a whole protein. This protein is broken in a high-energy system with warm air in a spectrometer, and the molecular mass data of the intact protein that avoided fragmentation, as well as the molecular weight data of the various pieces, is obtained. The information collected is then compared to existing protein sequences and fragments in a database to get the protein's entire amino acid pattern. An examination like that might reveal a disparity in the molecular mass of a particular fragment, which generally signals a protein change. It can determine which amino acid has been modified. (Han et al. 2006).

3.4 DETERMINATION OF THE 3D S TRUCTURE OF A PROTEIN

The amino acid pattern and three-dimensional shape of a protein may be determined by techniques apart from the mass spectrometric examination. Before spectroscopic protein analysis, the amino acid pattern is established using Edman degradation. Mass spectrometry has superseded the labor-intensive and time-consuming Edman degradation technique. Mass spectrometric analysis cannot be used to identify the three-dimensional structure of a protein. Nevertheless, the mass spectrometry-obtained amino acid sequence data may be utilized in conjunction with a protein dataset to predict or build the three-dimensional structure of certain proteins. X-ray crystallography (XRC), X-ray diffraction, and nuclear magnetic resonance (NMR) are the only direct methods for establishing the 3D shape of the protein. However, XRC and NMR each have their benefits and disadvantages, which will be explored in the following part.

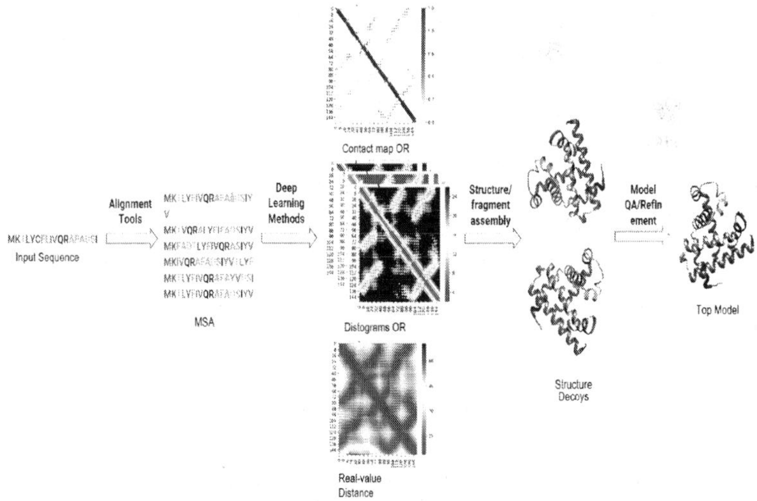

Figure 3.10. demonstrates that protein structure, amino acid makeup, and sequence influence proteome susceptibility to oxidation-induced degradation.

Source: Pakhrin SC, Shrestha B, Adhikari B, KC DB. Deep Learning-Based Advances in Protein Structure Prediction. International Journal of Molecular Sciences. 2021; 22(11):5553. https://doi.org/10.3390/ijms22115553

3.4.1 X-Ray Crystallography/X-Ray Diffraction

The crystals of a pure protein are subjected to an X-ray laser in just this procedure. The arrangement of atoms contained in a protein crystal diffracts the X-ray. The number of electrons in the particles and the structure of particles in a protein molecule determine the X-ray diffraction sequence. An electron network is designed by diffraction pattern X-rays that are received as a reflection on a sensor. That map is used to build a depiction of the atoms in a protein complex to show the protein's three-dimensional structure. X-ray crystallography has brought a new age in biology by elucidating the 3d image of various biomolecules and nucleotides. J. D. Bernal of Cambridge University discovered X-ray crystallography about 70 years ago when he produced the first 3D structure of a tiny protein called pepsin. In 1958, Max Perutz, a colleague of Bernal and Sir John Kendrew, established the three-dimensional structure of the very first major proteins (hemoglobin and myoglobin) using X-ray crystallography, about which they received the Nobel Prize in Chemistry. X-ray diffraction was also used to resolve the shape of the DNA duplex, which would have been done by Rosalind Franklin, a Bernal student working in Maurice Wilkins' group. Their findings of the double-helical shape of the DNA molecule were predicated on Rosalind Franklin's X-ray diffraction peak. In 1962, Maurice Wilkins, James Watson, and Francis Crick received the Nobel Prize in Medicine for elucidating the double-helical DNA structure.

X-ray crystallography is a time-consuming technique for determining a protein's 3D structure. Among the most challenging aspects of this process is obtaining pure protein and afterward creating crystals of protein complexes. Crystallized proteins are hard to come by. The production of protein crystals seems more like artwork as compared to science. It's challenging to crystalline low-abundance proteins, especially hydrophobic proteins or proteins having hydrophobic regions, like protein molecules. X-ray crystallography cannot be used to study such proteins. Nonetheless, the latest innovations have solved certain challenges with protein crystallization, such as the introduction of robots or crystal workstations that employ many factors concurrently to create optimal protein crystallization.

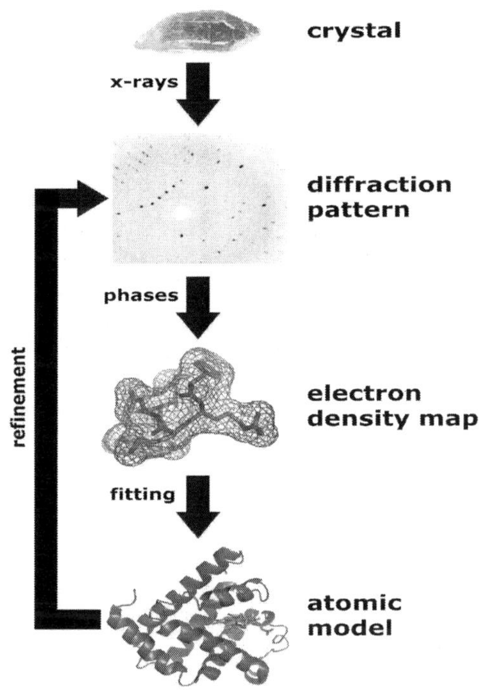

Figure 3.11. X-ray crystallography process for molecular structural characterization.

Source: https://commons.wikimedia.org/wiki/File:X_ray_diffraction.png

Aside from the difficulties of crystallization, X-ray crystallography creates a massive volume of data that should be evaluated to obtain a 3D image of the protein molecule. X-ray study of the crystalline protein molecule or nucleic acids wouldn't be conceivable with no utilization of computers and bioinformatics.

3.4.2 Neutron Scattering

Neutron diffraction also reveals the three-dimensional shape of the protein. A protein crystal is subjected to the neutron ray, and the dispersed neutron determines the location of molecules inside the protein. In contrast to X-ray diffraction, the neutron is dispersed by the nuclei of atoms and not by electrons; hence, this technique produces a unique type of actual nuclear image than X-ray crystallography. By nuclear reactions, a limited set of genes have been examined. There are just a few neutron bouncing instruments around the globe, which is a significant drawback of this technique. Using this technology, less than a few proteins have indeed been examined at the atomic scale.

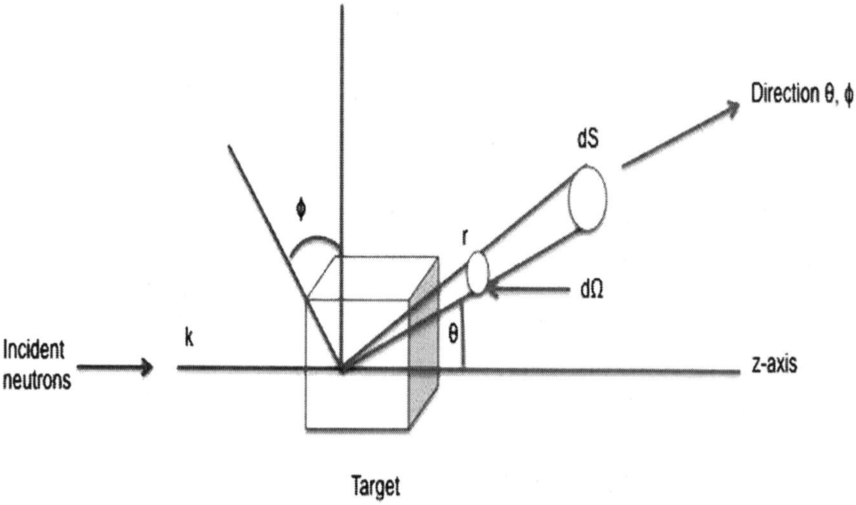

Figure 3.12. Graphic depiction of the dispersion of neutrons as they strike an object.

Source:https://chem.libretexts.org/Bookshelves/Analytical_Chemistry/Physical_Methods_in_Chemistry_and_Nano_Science_(Barron)/07%3A_Molecular_and_Solid_State_Structure/7.05%3A_Neutron_Diffraction

3.4.3 Nuclear Magnetic Resonance Spectroscopy

The magnetic characteristics of the nucleus of specific atoms provide the basis for NMR spectroscopy. Atoms with unusual protons in the nucleus or neutrons have spinning and act like magnets. As a result, hydrogens (H1) or some steady isotopes with just unusual protons in the nucleus or neutrons, like H2, C13, N15, P31, and F19, act like magnets. Such particles collect radio signals of the same rate as their rotation whenever subjected to radio signals of specific frequencies in an externally applied magnetic field; this technique is termed "resonance." This change in the atom's condition is known as "chemical shift," which is used to identify the atom's chemical characteristics. Following the reception of radio signals, the atoms get energized, but subsequently, produce radiation equivalent to the quantity received. The quantity of radiation released as well as the time required to release it may both be monitored and utilized to learn about the structure of the atom (Spáčil et al., 2008).

NMR was invented separately by Felix Bloch at Stanford University and Edward Purcell at MIT, who both received the Nobel Prize in Physics in 1952. Whilst the NMR signal supplied certain data around an atom's nucleus, it had a limited resolution and a limited signal-to-noise (S/N) ratio. Robert Ernst used Discrete Fourier transform to solve this issue of NMR, and 2D and multidimensional NMR were soon accessible; for such achievements, Robert Ernst earned the Nobel Prize in Chemistry in 1991. Kurt Wuthrich later invented NMR technology that could be used to analyze the three-dimensional shape of a protein complex, by which he won a Nobel Prize in Chemistry in 2002. (Wuthrich 2002).

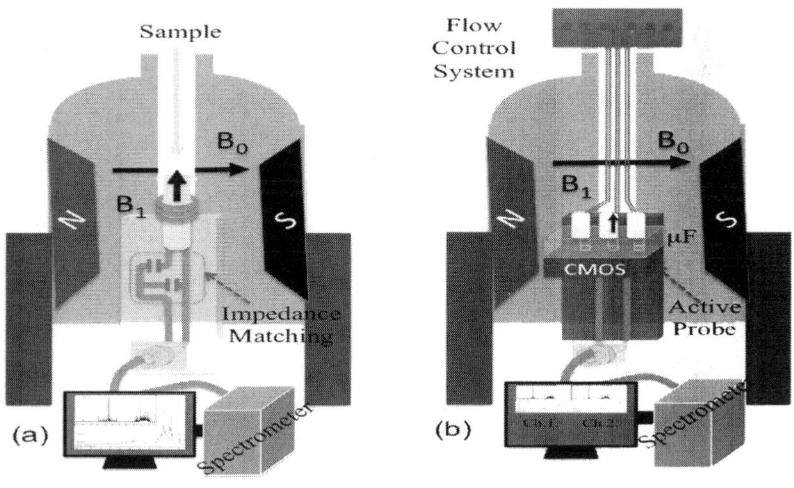

Figure 3.13. Nuclear Magnetic Resonance (NMR) Spectroscopy Theory.

Source; Pourmodheji H, Ghafar-Zadeh E, Magierowski S. A Multidisciplinary Approach to High Throughput Nuclear Magnetic Resonance Spectroscopy. Sensors. 2016; 16(6):850. https://doi.org/10.3390/s16060850

The most prevalent component in a protein is hydrogen (H1). The 3D sequence of a protein is deciphered using NMR evidence for H1. The NMR values for H1 vary depending on their location; for instance, the NMR signal for H in CH3 differs from CH or OH. H connected to C also has a distinct NMR signal than H linked to N. Furthermore, whenever two neighboring H atoms bonded by bonds are formed escape as non - covalently linked or as individual H atoms occurring separately or far from each other, they produce distinct NMR signals. Since multifunctional NMR spectroscopy is now available, now it is possible to identify the many types of NMR signals for H1. In essence, three distinct types of NMR signals are produced. Nuclear Overhauser effect spectroscopy (NOESY), correlation spectroscopy (COSY), and total correlation spectroscopy (TOCSY) are examples of these techniques. The Nuclear Overhauser effect refers to NMR results from atoms that are near in dimension but not connected by covalent connections. This approach is effective for determining the locations of atoms in a protein structure when numerous faraway elements are pulled with each other in space due to polypeptide section bending, presuming the protein's secondary structure. COSY is a method for determining the locations of atoms connected by chemical reactions. TOCSY is a method of identifying atoms that really are sections of a network but are not linked by chemical bonds. All of these data are utilized to determine the ranges between individual atoms and to decode the protein's 3D structure. In terms of getting data on the positions of hydrogen atoms, additional data about carbon and nitrogen atoms is required. To do so, proteins are tagged with C13 and/or N15 and then expressed in bacteria utilizing the cloned gene. NMR is used to study the labeled proteins and determine where the C and N atoms are located in the protein (Horvath et al., 1976).

3.5 DETERMINATION OF THE NUMBER OF PROTEINS

Various approaches for identifying proteins and determining their primary, secondary, and three-dimensional structures to comprehend their involvement in proteomics have been described. In proteomics, though, it is crucial to learn the number of distinct proteins with their structural characteristics. There are several methods for determining the number of proteins in a specimen. The majority of such procedures are colorimetric, like the Lowry or Bradford's tests, and the spectrophotometric technique, which calculates the protein concentration dependent on the ultraviolet light at 280 nm. Such techniques compute the overall protein content of a specimen. Such techniques are incapable of determining the proportions of e various proteins in a given sample. Consequently, such approaches do not apply to the high-throughput analysis necessary in proteomics. Since this proteome fluctuates with its processes of the cell and in reaction to environmental events, it is essential to understand the abundances of various proteins in proteomics. In multicellular animals, the protein composition of each cell type varies, as do the growth requirements of a specific cell type. Furthermore, human cells and diseased cells may vary in their protein composition. Secondly, techniques for quantifying the various proteins in a specimen are explored. Procedures rely on antigen-antibody associations, like radioimmunoassay (RIA) in a solution or Western analyses, that analyzes the antigen-antibody association on a gel, may be used to assess the relative abundance of specific proteins. The radioimmunoassay was created by Rosalyn Yalow, who'd been awarded the Nobel Prize in Medicine and Physiology in 1977 for her contribution to the identification of insulin in human blood. She has been the second lady to win the Nobel Prize in Medicine and Physiology in mankind. Her RIA work cleared the path for the measurement of minute levels of proteins and hormones in the human body. Proteomics necessitates the determination of the high abundance of a huge range of proteins in representative groups, which is not possible with current techniques. Contrary to the goals of proteomics, RIA and Western testing may offer insight into the relative quantity of a single protein at a moment (MESBAH et al., 1989).

3.5.1 Quantitation of Proteins After Separation on a 2D Gel

A range of methodologies may be used to determine the number of proteins that occur in a huge number band segregated on a 2D gel, as well as their abundances. The strength of protein bands is assessed by the brightness of dyes employed to see the bands after the various protein specimens are resolved on a 2D gel. A densitometric scan may be used to determine the magnitude of protein bands. Silver staining is a sensitive process. Staining with a fluorescent dye is also effective. Developing cells in the existence of radiological amino acids, like methionine carrying S35 sulfur, labels proteins in two kinds of cells. A 2D gel is used to isolate the protein specimen from the two kinds of cells. The protein bands are visible on the gel as dots on an X-ray picture. Densitometry determines the strength of every point on the image.

3.5.2 Differential Gel Analysis

Therefore in the procedure, protein molecules from two distinct cell kinds are acquired, with one specimen stained with Cy3 flour (providing a red color) and the second specimen stained with Cy5 flour (producing a green color) (releasing green color). The specimens are combined and then run together on a 2D gel. Visualization of protein bands upon separation. A specific band having an equal quantity of proteins with Cy3 and Cy5 flours appears as a yellow point. Nevertheless, if there is a significant change in protein concentration within a band, these proteins will show as a green or red spot. A sample with an abundance of Cy3 flour seems red, while those with an abundance of Cy5 flour look green. Therefore, the variously colored bands indicated the proportional presence of various proteins (Šesták, et al., 2015).

3.5.3 Quantitative Spectrometry of Proteins

In a proteome study, spectrometry has been utilized to analyze the high protein content of specimens to estimate the frequency of numerous proteins at a similar time. The relative quantity of proteins in two specimens derived from two distinct kinds of cells may be measured in at minimum two methods. The first cell type is produced in ordinary water having hydrogen, while the second cell type is grown in water that contains deuterium. Proteins from either kind of cell are isolated individually, then combined, and evaluated in a spectrometer. Every protein shows as a doublet on the spectrometric graph in an assessment: Since their m/z ratios vary, one indicates the protein has hydrogen and the other indicates the protein has deuterium. The proportion of such graph heights will be used to determine their abundances (Horvath & Melander, 1977).

The second principle, like the first, is dependent on tagging proteins differently with hydrogen and deuterium. Every protein sample, though, is given isotope-coded affinity tags (ICATs). This method includes processing protein specimens with iodoacetamide biotin derivatives, which bind with cysteine residues in proteins. Such proteins are subsequently isolated using affinity chromatography on a column that contains streptavidin and binds biotin specifically. Such proteins are then isolated, combined, and spectrometer-analyzed. The spectrometer produces a doublet of proteins having hydrogen and deuterium once again. Their abundance is determined by measuring their elevation in a doublet (Ishii et al., 1977).

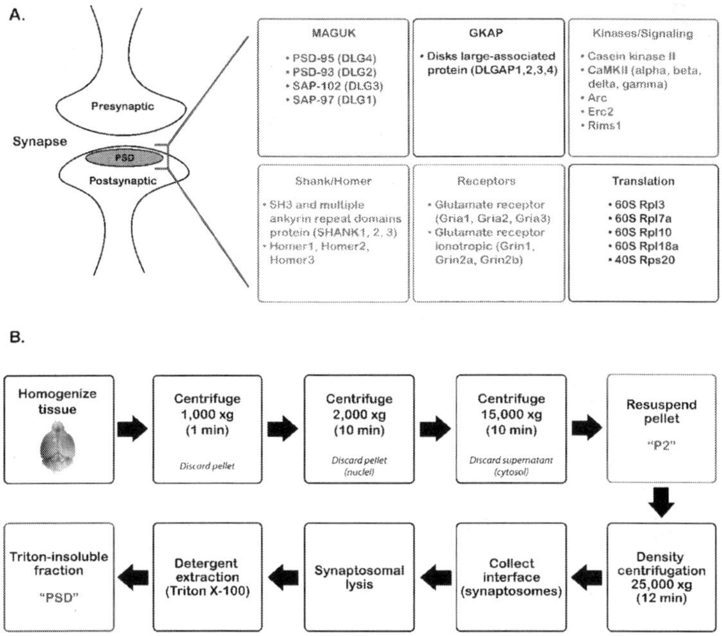

Figure 3.14. General-workflow-for-quantitative-mass-spectrometry-based-translational-neuroproteomics.

Source: Wilson RS, Rauniyar N, Sakaue F, Lam TT, Williams KR, Nairn AC. Development of Targeted Mass Spectrometry-Based Approaches for Quantitation of Proteins Enriched in the Postsynaptic Density (PSD). Proteomes. 2019; 7(2):12. https://doi.org/10.3390/proteomes7020012

3.5.4 Protein Microarray

Protein microarrays have helped evaluate the frequency of a large number of proteins across two cell types produced beneath two different environments. This method is ideal for proteome research because it allows researchers to track changes in a large number of proteins throughout two kinds of cells at the same time (Xiao &Oefner, 2001).

A microarray is a protein sample in which several antibodies are placed in specific locations on a microscope glass plate. The slide is then submerged in a solution that contains. The dots representing protein-protein interactions are detected and used to detect a wide range of proteins and their amounts on a specified scale. By comparing the brightness of markings on several slides subjected to protein soluble from several types of cells or by a type of cell cultivated in different conditions, the relative amount of individual proteins may be calculated. It might indicate an increase in a protein's activity in a variety of cells. In comparison to a similar microarray sheet of a cell line solution, the protein solution from malignant cells may exhibit a decline or spike in as many as several proteins. A protein solubility microarray from a kind of cell kept in a minimal or enriched medium will show changes in a variety of proteins, as well as the presence or absence of particular proteins. Microarrays are being used to identify not just the abundances of certain proteins, but also a variety of other associations, such as enzyme-substrate interactions and the discovery of a variety of other chemical compounds that interact with protein molecules or enzymes. A number of these microarray components will be discussed in the following chapters.

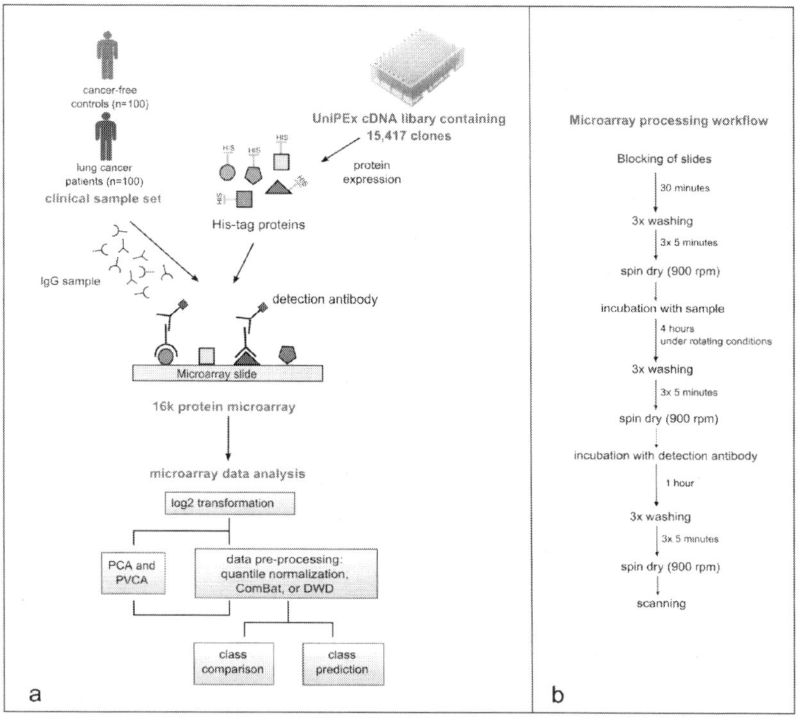

Figure 3.15. Protein-microarrays-a-Functional-protein-microarrays-for-studying-proteins.

Source: Brezina S, Soldo R, Kreuzhuber R, Hofer P, Gsur A, Weinhaeusel A. Immune-Signatures for Lung Cancer Diagnostics: Evaluation of Protein Microarray Data Normalization Strategies. Microarrays. 2015; 4(2):162-187. https://doi.org/10.3390/microarrays4020162

3.6 STRUCTURAL AND FUNCTIONAL PROTEOMICS

The basic structure is a protein's arrangement of amino acids, that defines its tertiary and secondary structures. Moreover, it gives the protein various elements and characteristics. The mass of the molecule, its pI, and its three-dimensional structure may be simply determined by the chain of amino acids. It is also possible to determine the activity of a protein based on its shape, for instance, if it operates like an enzyme or a sensor protein, as a hormonal or antibody, or as a protein known. It may also determine if it dwells in distinct sites inside a cell, is a component of the wall, or is released from the cell. The three-dimensional structure of a protein is dictated by how its amino acid chain folds around itself, putting essential amino acids into proximity. That folding structure identifies the enzymatic activity for binding or locations that interact with drugs, ligands, and inhibitors. Moreover, this could identify additional control areas of an enzyme protein on its interface or in the crevices of its three-dimensional structure. The correct packing of amino acids affects the total surface charge of a protein, resulting in a stable crystalline structure. The findings of an *in vitro* investigation involving the denaturation and re-denaturation of proteins support the notion that the amino acid pattern or the fundamental structure of a protein contains all the data necessary to determine its form and composition. Understanding the principles that regulate the folding of a protein into its natural 3D structure is crucial (Martin &Guiochon, 2005).

Even as an amino acid structure of a protein is stored in the nucleotide gene sequence, it is now feasible to identify the amino acid residues or basic shapes of a huge range of proteins by decoding the genomic regions of an organism that have been placed in the gene dataset. This has led to increased levels of the amino acid composition of a huge array of an organism's proteins. This is known as the PDB. The yeast PDB contains data on all 6000 proteins produced by the yeast genomes. The PDB of higher species is far from complete; the classification of extensive stretches of nucleotide sequences in people and other higher creatures into proteins with known features has to be decoded. Conventionally, the activity of a protein is defined by its biochemical analysis, often by its capacity to activate a biochemical process. This laborious task involves purifying a protein and then identifying the biological process that it catalyzes. This cannot be done on a huge scale. Consequently, going to follow a throughput strategy in proteomics, the feature of a protein is calculated by evaluating the basic structure and/or the three-dimensional framework of the protein in inquiry with the main sequence and/or the three-dimensional framework of a huge range of proteins obtainable in the PDB. Following is a discussion of the numerous facets of the link connecting protein structure and function (Guillarme et al., 2010).

3.6.1 Moonlighting by Protein

So we can go into detail about the link between a protein's function and structure, it's important to note that most proteins have just one function. This viewpoint is based on Beadle and Tatum's previous idea of the one-gene–one-enzyme hypothesis (1941). According to this view, each protein has just one activity, or one gene-one enzyme–one function. In larger organisms, we have previously shown that the one gene-one enzymatic model is invalid. And over one gene changes the development of nearly half of all proteins in humans. Similarly, it is now known that a protein might influence many functions (Yarnell, 2003). A vast number of proteins are now recognized to be versatile. Proteins such as the tumor suppressor P53 and the Warner syndrome protein WRN, which has multiple helicases and nuclease activities, are instances of multifunctional proteins. Distinct patterns or regions are found in all multipurpose proteins, every directing diverse functions. As a result, it promotes the notion of exon reshuffling for gene formation, which will be described later. Different variables can influence the diverse activities of a similar protein. Phosphoglucoisomerase (PGI), for instance, stimulates a metabolic step in the cell's internal glycolysis. Whenever released to the exterior of the cell, the same enzyme functions as cytokine neuroleukin, which helps B cells mature into antibiotic cells. That protein also has a neural growth parameter function. Whenever coupled to a specific receptor, the blood-clotting molecule thrombin acts as a cytokine. A protein's function can also change depending on where it is found in the cell. Whenever the bacterial protein Put A is linked with the cell surface, it operates as an enzymatic dehydrogenase, so when it's loose inside the cell, it acts as a transcriptional activation that wraps to DNA and regulates transcription. A protein's activity may vary across cells. Neuropilin, for instance, affects the formation of new blood cells in endothelial cells but not the orientation of brain development in an axion. Also, a protein's role could vary based on if it occurs as a monomer or a multimer inside a cell; for instance, the similar protein operations as glyceraldehyde-3-phosphate dehydrogenase in the glycolysis when found in a tetrameric type, but then as uracil-DNA glucosylase in the DNA repair path when found in a monomeric form. Moonlighting refers to a protein's ability to perform many functions. (Yarnell, 2003).

3.6.2 Determining the Primary Structure of Proteins by Different Methods

As discussed above, there are three ways for showing the order of amino acids in a protein to identify its main structure. (a) decoding the amino acid composition from the nucleotide pattern of DNA or cDNA; (b) Edman degradation; and (c) mass spectrometry. The amino acid pattern of a protein is easily extracted from its corresponding DNA sequence. Nevertheless, in larger species using splicing processes to produce messenger RNA (mRNA) for protein synthesis, it is important to detect the nucleotides at the intersection of exon and intron. Under the existence of conserved nucleotides, exon-intron junctions may be easily detected. Consequently, the DNA coding sequence is determined by recognizing and disregarding the intervening noncoding intron regions. Typically, the constantly coding DNA sequence of a gene is derived from the target DNA of complementary DNA (cDNA), which is produced by replicating the mRNA for a certain protein. The data sets from the sample of DNA of a species and the cDNA library of tissues of a species are thus valuable for interpreting the amino acid sequences of various proteins. Such information is now freely accessible in the gene and protein databases (Linden et al., 1975; Blakley& Vestal, 1983).

3.6.2.1 Determination of Amino Acid Sequence by Edman Degradation

This approach may be used to figure out the pattern of amino acids of a polypeptide with up to 50 amino acids. Typically, fragmentation of a larger peptide exposed to moderate hydrolysis or breakage by endopeptidases produces numerous tinier overlapping peptides. Such finer particles are analyzed, and the ordering of amino acids is determined by looking for matching amino acids in the pattern of a smaller peptide fragmented from a bigger peptide.

Before performing an amino acid analysis, it is essential to identify the overall amino acid composition of a peptide. This aids in determining the peptide's ultimate amino acid sequence. Knowing composition also aids in determining if the peptide has two contiguous glycine residues (mol wt 57 daltons) residues or only one asparagine residue (mol wt 114 daltons). This conundrum is founded only on molecular weight data.

Thorough hydrolysis of the peptide with 6 M HCl at a high temperature determines the entire amino acid makeup. To extract the amino acids, the hydrolysate is passed through paper electrophoresis. A dye named dinitrophenol is used to identify the various amino acids. Except for proline, which looks blue, all amino acids create yellow dots. The quantity of every amino acid in a given location is determined by the color intensity. In electrophoresis, the individual amino acid spots are recognized by matching their motility to recognized amino acids for use as standards (Sawardeker et al., 1965; Dugo et al., 2008).

Edman degradation determines the sequence of amino acids in a peptide. As previously stated, a larger peptide is broken into smaller peptides with overlaps. The sequence of amino acids in the peptide is determined by the overlapped amino acid(s). A peptide of three alanine, one arginine, two glutamic acids, two lysines, one phenylalanine, and one threonine, for instance, maybe split into smaller fragments. These have been proven to get the corresponding patterns using the technique of Edman degradation organized in an overlapping manner:

lysgluthrala

alaalaalalysthralaala

alalysphenglu

phengluarg

The structure of such 10 amino acids in the peptide may thus be derived as follows:lysgluthralaalaalalysphengluarg.

Chemical cleavage using cyanogen bromide or enzymatic breakage using endopeptidases like trypsin or chymotrypsin are used to break bigger peptides (Lindsey et al., 2006; Song et al., 2010).

Just before doing Edman degradation sequencing, it is usually beneficial to know the N-terminal amino acid in the peptide. The N-terminal amino acid is detected by combining the peptide using Sanger's reagent fluorodinitrobenzene (FDNB), separating the amino acids on electrophoresis, and identifying the N-terminal amino acid by the color produced by Sanger's reagent reactions. The color of FDBN is yellow. Dansyl chloride, other than FDBN, could be used to tag the N-terminal amino acid. Like this in an examination, the existence of more than one N-terminal amino acid indicates the existence of much more than one peptide in the protein, as was discovered in the case of insulin. The amino acid pattern of the bigger component may sometimes be determined by examining the sequence of segments created by two separate enzymes. For instance, Edman degradation revealed that tryptic digestion of a peptide produces two short breaks with the following amino acid composition:

Ala-alatrpglylys;thrasnvallys

Nevertheless, the identical peptide is digested by a shorter peptide with the pattern shown below:

vallysalaalatrp.

A study of such three patterns reveals that the bigger peptide's amino acid pattern must be thrasnvallys ala alatrpglylys, as shown by Stryer(1982).

3.6.2.2 Determination of AminoAcid Sequence by Mass Spectrometry

Typically, the amino acid structure of a protein is established by mass spectrometric analyses of the peptides and their components. The rapidity and scope of spectrometric analysis are compatible with the goals of proteome investigations. A 2D gel-separated protein is typically injected into a mass spectrometer with two quadrupole analyzers and a TOF analyzer. The broken ionized particles approach the very first quadrupole analyzer. A specific peptide segment is preferentially permitted to approach the second quadrupole detector, in which it is combined with neutral gases like nitrogen as well as argon. The gases aid in separating the peptide fragment's peptide bonds. To create tiny parts with lesser amino acids per fragmentation, care is designed to fail the protein complex even once in every fraction, although at many sites, to minimize the number of amino acids per segment. Alternately, it is feasible to delete one amino acid at a moment from the C-terminus ending to form an arrangement of peptides that seems to be one amino acid smaller than the subsequent larger chain in the sequence. Every time this fragmentation happens, a charged fragment with an N-terminal end and an uncharged fragment with such a C-terminus are produced. The charged particles pass via the TOF detector en route to the sensor. Their flight duration is proportional to their mass; tiny particles pass quicker than bigger ones. Thus, the time required to reach the sensor is precisely proportional to the molecule's mass. On the lateral x-x axis of a chart, the flight duration is noted; the height on the y-y axis shows the concentration or quantity of each fragment. A typical finding relates the weight of the various segments of an 8-amino acid chain peptide (Wagner et al., 2003; Grinias et al., 2016).

3.6.2.3 Determining the Secondary Structure of Protein

The fundamental basis of a peptide, or its amino acid sequence, determines its secondary and tertiary structure. Depending upon the denaturation or unraveling of an enzyme ribonuclease in the presence of urea and the folded form of the similar enzyme with complete elimination of the denaturing agent, urea, Christian Anfinsen discovered this finding. Knowing the secondary structure of the protein is necessary before learning about the tertiary expression of proteins. Since of their significance in cell processes and modification in biotechnology and medication synthesis, it's critical to understand the principles which proteins take to adopt a 3D shape (Dunham et al., 2011).

Since contact between the main chain of neighboring amino acids, the linear peptide begins to fold upon itself its original shape. This results in the creation of various configurations, including (a) an alpha helix, (b) stranded folds known as beta sheets or beta strands, and (c) random coils. Some parameters may now be utilized to anticipate the presence of such components in the protein's secondary structure. Chow and Fasman (1978) were the first to discover this, depending on the affinity of particular amino acids to connect with such structures, namely the helix and beta-sheet. The amino acids Glu, met, ala, and lys, for example, are mostly related to the helix structure, while val, ile, and tyr are mostly connected with the beta-sheet structure. Either helix or the beta-sheet are connected with the amino acid leucine. The amino acids glycine and proline act as helix breakers, with proline being the first residue in the helix. At the N-terminus, asp and glu are found, while arg and lys are found at the C-terminus. These are some generalizations based on Chou and Fasman's statistical study (also see Chou and Kai, 2004).

The presence of hydrophobic amino acids grouped on the top of globular proteins seems to be another feature of protein secondary structure. Hydrophobic amino acids appear to be the twentieth amino acid in the peptide's main building regularly. Such amino acids are twisted on the axes of the protein by 100 degrees, resulting in globular proteins with all hydrophobic amino acids grouped on one end on the helical protein surface and polar amino acids at another end. As a result, a wheel of amino acids may be generated in a protein, with hydrophobic amino acids grouped with one face of the helix and polar amino acids on another.

The structure of proteins in the proteins dataset may now be examined using a variety of software packages. Furthermore, computer simulations are being created to see the secondary and tertiary structure of proteins, that occurs within microseconds after the main structure's development. Such computer simulations are likely to provide researchers with a better understanding of protein structure, as well as the principles for secondary and tertiary structure synthesis (Stobaugh et al., 2013; Unger et al., 2000).

REFERENCES

1. Abdelhamid, H. N., & Wu, H. F. (2015). Proteomics analysis of the mode of antibacterial action of nanoparticles and their interactions with proteins. *TrAC Trends in Analytical Chemistry*, *65*, 30-46.
2. Acquah, C., Chan, Y. W., Pan, S., Agyei, D., &Udenigwe, C. C. (2019). Structure-informed separation of bioactive peptides. *Journal of Food Biochemistry*, *43*(1), e12765.
3. Alharbi, R. A. (2020). Proteomics approach and techniques in identification of reliable biomarkers for diseases. *Saudi journal of biological sciences*, *27*(3), 968-974.
4. Arima, Y., & Iwata, H. (2007). Effect of wettability and surface functional groups on protein adsorption and cell adhesion using well-defined mixed self-assembled monolayers. *Biomaterials*, *28*(20), 3074-3082.
5. Baker, D. and A., Sali. 2001. Protein structure prediction and structural genomics. *Science* 294, 93–96.
6. Beadle, G. W. and Tatum, E. L. 1941. Genetic control of biochemical reactions in Neurospora. Proc. Natl. Acad. Sci. U.S.A. 27, 499–506.
7. Bellot, J. C., &Condoret, J. S. (1993). Modelling of liquid chromatography equilibria. *Process Biochemistry*, *28*(6), 365-376.
8. Bergström, S. K., Dahlin, A. P., Ramström, M., Andersson, M., Markides, K. E., &Bergquist, J. (2006). A simplified multidimensional approach for analysis of complex biological samples: on-line LC-CE-MS. *Analyst*, *131*(7), 791-798.
9. Blakley, C. R., & Vestal, M. L. (1983). Thermospray interface for liquid chromatography/mass spectrometry. *Analytical Chemistry*, *55*(4), 750-754.
10. Bowie, J. U., R. L¨uthy, and D. Eisenberg. 1991. A method to identify protein sequences that fold into a known three-dimensional structure. *Science* 253, 164–170.
11. Brohee, S., & Van Helden, J. (2006). Evaluation of clustering algorithms for protein-protein interaction networks. *BMC Bioinformatics*, *7*(1), 1-19.
12. Büyükköroğlu, G., Dora, D. D., Özdemir, F., &Hızel, C. (2018). Techniques for protein analysis. In *Omics technologies and bio-engineering* (pp. 317-351). Academic Press.

13. Capriotti, A. L., Cavaliere, C., Foglia, P., Samperi, R., &Laganà, A. (2011). Intact protein separation by chromatographic and/or electrophoretic techniques for top-down proteomics. *Journal of Chromatography A, 1218*(49), 8760-8776.
14. Cassidy, L., Helbig, A. O., Kaulich, P. T., Weidenbach, K., Schmitz, R. A., &Tholey, A. (2021). Multidimensional separation schemes enhance the identification and molecular characterization of low molecular weight proteomes and short open reading frame-encoded peptides in top-down proteomics. *Journal of Proteomics, 230*, 103988.
15. Cedervall, T., Lynch, I., Lindman, S., Berggård, T., Thulin, E., Nilsson, H., ... &Linse, S. (2007). Understanding the nanoparticle–protein corona using methods to quantify exchange rates and affinities of proteins for nanoparticles. *Proceedings of the National Academy of Sciences, 104*(7), 2050-2055.
16. Chou, K. C. and Y. D. Kai. 2004. A novel approach to predict active sites of enzyme molecules. *Proteins* 55, 77–82.
17. Chou, P. Y. and C. D. Fasman. 1978. Empirical prediction of protein conformation. *Ann. Rev. Biochem*. 47, 251–278.
18. Chouchani, E. T., James, A. M., Fearnley, I. M., Lilley, K. S., & Murphy, M. P. (2011). Proteomic approaches to the characterization of protein thiol modification. *Current opinion in chemical biology, 15*(1), 120-128.
19. Comisarow, M. B. and A. G. Marshall. 1974. Fourier transformion cyclotron resonance spectroscopy. Chem. Phys. Lett. 25, 282–283.
20. Covey, T. R., Lee, E. D., Bruins, A. P., &Henion, J. D. (1986). Liquid chromatography/mass spectrometry. *Analytical chemistry, 58*(14), 1451A-1461A.
21. Di Palma, S., Hennrich, M. L., Heck, A. J., & Mohammed, S. (2012). Recent advances in peptide separation by multidimensional liquid chromatography for proteome analysis. *Journal of proteomics, 75*(13), 3791-3813.
22. Dixon, S. P., Pitfield, I. D., & Perrett, D. (2006). Comprehensive multi-dimensional liquid chromatographic separation in biomedical and pharmaceutical analysis: a review. *Biomedical Chromatography, 20*(6-7), 508-529.

23. Dugo, P., Cacciola, F., Kumm, T., Dugo, G., &Mondello, L. (2008). Comprehensive multidimensional liquid chromatography: theory and applications. *Journal of Chromatography A, 1184*(1-2), 353-368.

24. Dunham, W. H., Larsen, B., Tate, S., Badillo, B. G., Goudreault, M., Tehami, Y., ... &Gingras, A. C. (2011). A cost–benefit analysis of multidimensional fractionation of affinity purification-mass spectrometry samples. *Proteomics, 11*(13), 2603-2612.

25. Fenn, J. B. M. Mann, G. K. Meng, S. F. Wong, and C. M. Whitehouse. 1989. Electrospray ionization for mass spectrometry of large biomolecules. *Science* 246, 64.

26. Fenn, John. B. 2002. Electrospray wings for molecular elephants. *Les Prix Nobel. The Nobel Prizes 2002* , Editor Tore Fr¨angsmyr, [Nobel Foundation], Stockholm,2003.

27. Field, J. K., Euerby, M. R., &Petersson, P. (2020). Investigation into reversed phase chromatography peptide separation systems part III: Establishing a column characterisation database. *Journal of Chromatography A, 1622*, 461093.

28. Fournier, M. L., Gilmore, J. M., Martin-Brown, S. A., & Washburn, M. P. (2007). Multidimensional separations-based shotgun proteomics. *Chemical reviews, 107*(8), 3654-3686.

29. François, I., Sandra, K., & Sandra, P. (2009). Comprehensive liquid chromatography: fundamental aspects and practical considerations—a review. *Analytica Chimica Acta, 641*(1-2), 14-31.

30. Gordon, S. M., Deng, J., Tomann, A. B., Shah, A. S., Lu, L. J., & Davidson, W. S. (2013). Multi-dimensional co-separation analysis reveals protein–protein interactions defining plasma lipoprotein subspecies. *Molecular & Cellular Proteomics, 12*(11), 3123-3134.

31. Gorg, A. 2000. The current state of two-dimensional electrophoresis with immobilized pH gradients. Electrophoresis 21, 1037–1053.

32. Gorg, A. et al. 1988. The current state of two-dimensional electrophoresis with immobilized pH gradients. Electrophoresis 9: 531–546.

33. Greibrokk, T., & Andersen, T. (2003). High-temperature liquid chromatography. *Journal of Chromatography A, 1000*(1-2), 743-755.

34. Grinias, K. M., Godinho, J. M., Franklin, E. G., Stobaugh, J. T., & Jorgenson, J. W. (2016). Development of a 45 kpsi ultrahigh pressure liquid chromatography instrument for gradient separations of peptides

using long microcapillary columns and sub-2 μm particles. *Journal of Chromatography A*, *1469*, 60-67.

35. Guillarme, D., Ruta, J., Rudaz, S., &Veuthey, J. L. (2010). New trends in fast and high-resolution liquid chromatography: a critical comparison of existing approaches. *Analytical and bioanalytical chemistry*, *397*(3), 1069-1082.
36. Hamilton, R. J., & Sewell, P. A. (1982). Introduction to high performance liquid chromatography. In *Introduction to high performance liquid chromatography* (pp. 1-12). Springer, Dordrecht.
37. Han, X., M. Jin, K. Breuker, and F. W. McLafferty. 2006. Extending topdown mass spectrometry to proteins with masses >200 kDa. Science 314, 109–112.
38. Haque, A. (1998). *Part I. Direct injection of selected medications in serum using restricted access media columns. Part II. Capillary electrophoretic measurements of selected drugs in pharmaceutical dosage forms. Part III. Enantiomeric separations of medicinals using a protein-bonded chiral stationary phase*. University of Georgia.
39. Horvath, C., &Melander, W. (1977). Liquid chromatography with hydrocarbonaceous bonded phases; theory and practice of reversed phase chromatography. *Journal of Chromatographic Science*, *15*(9), 393-404.
40. Horvath, C., Melander, W., & Molnar, I. (1976). Solvophobic interactions in liquid chromatography with nonpolar stationary phases. *Journal of Chromatography A*, *125*(1), 129-156.
41. Ishii, D., Asai, K., Hibi, K., Jonokuchi, T., & Nagaya, M. (1977). A study of micro-high-performance liquid chromatography: I. Development of technique for miniaturization of high-performance liquid chromatography. *Journal of Chromatography A*, *144*(2), 157-168.
42. Issaq, H. J. (2003). Application of separation technologies to proteomics research. *Advances in protein chemistry*, *65*, 249-269.
43. Issaq, H. J., Chan, K. C., Blonder, J., Ye, X., &Veenstra, T. D. (2009). Separation, detection and quantitation of peptides by liquid chromatography and capillary electrochromatography. *Journal of Chromatography A*, *1216*(10), 1825-1837.

44. Issaq, H. J., Chan, K. C., Janini, G. M., Conrads, T. P., &Veenstra, T. D. (2005). Multidimensional separation of peptides for effective proteomic analysis. *Journal of Chromatography B*, *817*(1), 35-47.

45. Issaq, H. J., Conrads, T. P., Janini, G. M., &Veenstra, T. D. (2002). Methods for fractionation, separation and profiling of proteins and peptides. *Electrophoresis*, *23*(17), 3048-3061.

46. Jacobs, J. M., Mottaz, H. M., Yu, L. R., Anderson, D. J., Moore, R. J., Chen, W. N. U., ... & Smith, R. D. (2004). Multidimensional proteome analysis of human mammary epithelial cells. *Journal of proteome research*, *3*(1), 68-75.

47. Jones, D. T., and C. Hadley. 2000. Threading methods for protein structure prediction. In D. Higgins, and W. R. Taylor (eds.) Bioinformatics: Sequence, Structure and Databanks. Heidelberg, Germany Springer-Verlag.

48. Jones, D. T., W. R. Taylor, and J. M. Thornton. 1992. A new approach to protein fold recognition. *Nature* 358, 86–89.

49. Jorrin-Novo, J. V. (2014). Plant proteomics methods and protocols. In *Plant proteomics* (pp. 3-13). Humana Press, Totowa, NJ.

50. Karas, M. and F. Hillenkamp. 1988. Laser desorption ionization of proteins with molecular masses exceeding 10,000 daltons. *Anal. Chem.* 60, 2299–2301.

51. Kay, B. K., Williamson, M. P., &Sudol, M. (2000). The importance of being proline: the interaction of proline-rich motifs in signaling proteins with their cognate domains. *The FASEB journal*, *14*(2), 231-241.

52. Klose, J. 1975. Protein mapping by combined isoelectric focusing and electrophoresis in mouse tissues. A novel approach to testing for induced point mutations in mammals. *Humangenetik* 26, 231–243.

53. Knox, J. H., & Grant, I. H. (1987). Miniaturisation in pressure and electroendosmotically driven liquid chromatography: Some theoretical considerations. *Chromatographia*, *24*(1), 135-143.

54. Kollman, P., I. et al.. 2000. Calculating structures and free energies of complex molecules: combining molecular mechanics and continuum models. *Acc. Chem. Res.* 33, 889–897.

55. Linden, J. C., &Lawhead, C. L. (1975). Liquid chromatography of saccharides. *Journal of chromatography A*, *105*(1), 125-133.

56. Lindsey, M. L., Goshorn, D. K., Comte-Walters, S., Hendrick, J. W., Hapke, E., Zile, M. R., &Schey, K. (2006). A multidimensional proteomic approach to identify hypertrophy-associated proteins. *Proteomics*, *6*(7), 2225-2235.
57. Link, A. J. (2002). Multidimensional peptide separations in proteomics. *Trends in Biotechnology*, *20*(12), s8-s13.
58. Mahmoudi, M., Lynch, I., Ejtehadi, M. R., Monopoli, M. P., Bombelli, F. B., & Laurent, S. (2011). Protein− nanoparticle interactions: opportunities and challenges. *Chemical reviews*, *111*(9), 5610-5637.
59. Martin, M., &Guiochon, G. (2005). Effects of high pressure in liquid chromatography. *Journal of Chromatography A*, *1090*(1-2), 16-38.
60. McCall, J. M. (1975). Liquid-liquid partition coefficients by high-pressure liquid chromatography. *Journal of medicinal chemistry*, *18*(6), 549-552.
61. McNeff, C. V., Yan, B., Stoll, D. R., & Henry, R. A. (2007). Practice and theory of high temperature liquid chromatography. *Journal of separation science*, *30*(11), 1672-1685.
62. MESBAH, M., PREMACHANDRAN, U., & WHITMAN, W. B. (1989). Precise measurement of the G+ C content of deoxyribonucleic acid by high-performance liquid chromatography. *International Journal of Systematic and Evolutionary Microbiology*, *39*(2), 159-167.
63. Meyer, B., & Peters, T. (2003). NMR spectroscopy techniques for screening and identifying ligand binding to protein receptors. *AngewandteChemie International Edition*, *42*(8), 864-890.
64. Nawrocki, J. (1997). The silanol group and its role in liquid chromatography. *Journal of Chromatography A*, *779*(1-2), 29-71.
65. Neverova, I., & Van Eyk, J. E. (2005). Role of chromatographic techniques in proteomic analysis. *Journal of Chromatography B*, *815*(1-2), 51-63.
66. Nguen, H. D. and K. Carol, Hall. 2004. Molecular dynamics simulation of spontaneous formation by random-coil peptides. Proc. Natl. Acad. Sci. U.S.A. 101, 16180–85.
67. Niessen, W. M. A. (1999). State-of-the-art in liquid chromatography–mass spectrometry. *Journal of Chromatography A*, *856*(1-2), 179-197.
68. Nikolin, B., Imamović, B., Medanhodžić-Vuk, S., & Sober, M. (2004). High perfomance liquid chromatography in pharmaceutical analyses. *Bosnian journal of basic medical sciences*, *4*(2), 5.

69. Novotny, M. (1981). Microcolumns in liquid chromatography. *Analytical chemistry*, *53*(12), 1294A-1308A.
70. O'Farrell, P. H. 1975. High resolution two dimensional electrophoresis of proteins. J. Biol. Chem. 250, 4007–4021.
71. Oppenheimer, J. A. 1986. 1945 First nuclear blast. In R. Rhodes, (ed.) The Making of the Atomic Bomb. New York: Simon & Schuster.
72. Peng, J., Elias, J. E., Thoreen, C. C., Licklider, L. J., &Gygi, S. P. (2003). Evaluation of multidimensional chromatography coupled with tandem mass spectrometry (LC/LC− MS/MS) for large-scale protein analysis: the yeast proteome. *Journal of proteome research*, *2*(1), 43-50.
73. Peng, Y., Pallandre, A., Tran, N. T., &Taverna, M. (2008). Recent innovations in protein separation on microchips by electrophoretic methods. *Electrophoresis*, *29*(1), 157-178.
74. Pitt, J. J. (2009). Principles and applications of liquid chromatography-mass spectrometry in clinical biochemistry. *The Clinical Biochemist Reviews*, *30*(1), 19.
75. Reinders, J., Zahedi, R. P., Pfanner, N., Meisinger, C., &Sickmann, A. (2006). Toward the complete yeast mitochondrial proteome: multidimensional separation techniques for mitochondrial proteomics. *Journal of proteome research*, *5*(7), 1543-1554.
76. Salzer, U., Hunger, U., &Prohaska, R. (2008). Chapter Three Insights in the Organization and Dynamics of Erythrocyte Lipid Rafts. *Advances in Planar Lipid Bilayers and Liposomes*, *6*, 49-259.
77. Sawardeker, J. S., Sloneker, J. H., & Jeanes, A. (1965). Quantitative determination of monosaccharides as their alditol acetates by gas liquid chromatography. *Analytical Chemistry*, *37*(12), 1602-1604.
78. Šesták, J., Moravcová, D., &Kahle, V. (2015). Instrument platforms for nano liquid chromatography. *Journal of Chromatography A*, *1421*, 2-17.
79. Song, C., Ye, M., Han, G., Jiang, X., Wang, F., Yu, Z., ... & Zou, H. (2010). Reversed-phase-reversed-phase liquid chromatography approach with high orthogonality for multidimensional separation of phosphopeptides. *Analytical chemistry*, *82*(1), 53-56.
80. Sorci, M., Gu, M., Heldt, C. L., Grafeld, E., & Belfort, G. (2013). A multi-dimensional approach for fractionating proteins using charged membranes. *Biotechnology and Bioengineering*, *110*(6), 1704-1713.

81. Spáčil, Z., Nováková, L., &Solich, P. (2008). Analysis of phenolic compounds by high performance liquid chromatography and ultra performance liquid chromatography. *Talanta*, *76*(1), 189-199.
82. Stobaugh, J. T., Fague, K. M., & Jorgenson, J. W. (2013). Prefractionation of intact proteins by reversed-phase and anion-exchange chromatography for the differential proteomic analysis of Saccharomyces cerevisiae. *Journal of Proteome Research*, *12*(2), 626-636.
83. Stoffel, W., Chu, F., & Ahrens, E. H. (1959). Analysis of long-chain fatty acids by gas-liquid chromatography. *Analytical Chemistry*, *31*(2), 307-308.
84. Stryer, L. 1982. Biochemistry Second Edition, W.H. Freeman and Company, San Francisco, CA.
85. Tanaka, K. 2002. The origin of macromolecule ionization by laser irradiation. Nobel Lecture 197–217.
86. Tanaka, K., H. Waki, Y. Ido, S. Akita, Y. Yoshida, and T. Yoshida. 1988. Protein and polymer analysis uo to m/z100.000by laser ionization time-of-flight mass spectrometry. *Rapid Commun. Mass. Spectrom*. 2, 151–153.
87. Tapuhi, Y., Schmidt, D. E., Lindner, W., &Karger, B. L. (1981). Dansylation of amino acids for high-performance liquid chromatography analysis. *Analytical biochemistry*, *115*(1), 123-129.
88. Trufelli, H., Palma, P., Famiglini, G., &Cappiello, A. (2011). An overview of matrix effects in liquid chromatography–mass spectrometry. *Mass spectrometry reviews*, *30*(3), 491-509.
89. Turriziani, B., Kriegsheim, A. V., & Pennington, S. R. (2016). Protein-protein interaction detection via mass spectrometry-based proteomics. *Modern Proteomics–Sample Preparation, Analysis and Practical Applications*, 383-396.
90. Unger, K. K., Racaityte, K., Wagner, K., Miliotis, T., Edholm, L. E., Bischoff, R., & Marko-Varga, G. (2000). Is multidimensional high performance liquid chromatography (HPLC) an alternative in protein analysis to 2D gel electrophoresis?. *Journal of High Resolution Chromatography*, *23*(3), 259-265.
91. Urey, C., Weiss, V. U., Gondikas, A., von der Kammer, F., Hofmann, T., Marchetti-Deschmann, M., ... &Andersson, R. (2016). Combining gas-phase electrophoretic mobility molecular analysis (GEMMA), light

scattering, field flow fractionation and cryo electron microscopy in a multidimensional approach to characterize liposomal carrier vesicles. *International journal of pharmaceutics*, *513*(1-2), 309-318.

92. Vissers, J. P., Claessens, H. A., &Cramers, C. A. (1997). Microcolumn liquid chromatography: instrumentation, detection and applications. *Journal of Chromatography A*, *779*(1-2), 1-28.
93. Wagner, Y., Sickmann, A., Meyer, H. E., &Daum, G. (2003). Multidimensional nano-HPLC for analysis of protein complexes. *Journal of the American Society for Mass Spectrometry*, *14*(9), 1003-1011.
94. Washburn, M. P. (2004). Utilisation of proteomics datasets generated via multidimensional protein identification technology (MudPIT). *Briefings in Functional Genomics*, *3*(3), 280-286.
95. Wolters, D. A., Washburn, M. P., & Yates, J. R. (2001). An automated multidimensional protein identification technology for shotgun proteomics. *Analytical chemistry*, *73*(23), 5683-5690.
96. Wren, S. A., &Tchelitcheff, P. (2006). Use of ultra-performance liquid chromatography in pharmaceutical development. *Journal of Chromatography A*, *1119*(1-2), 140-146.
97. Wuthrich, K. 1986. NMR of Proteins and Nucleic Acids. New York: Wiley.
98. Wuthrich, K. 2002. NMR studies of structure and function of biological macromolecules. Nobel Lecture 235–267.
99. Xia, S., Tao, D., Yuan, H., Zhou, Y., Liang, Z., Zhang, L., & Zhang, Y. (2012). Nano-flow multidimensional liquid chromatography platform integrated with combination of protein and peptide separation for proteome analysis. *Journal of separation science*, *35*(14), 1764-1770.
100. Xiao, W., &Oefner, P. J. (2001). Denaturing high-performance liquid chromatography: A review. *Human mutation*, *17*(6), 439-474.
101. Yarnell, A. 2003. The double lives of enzymes. C&EN 81, 35–36.
102. Zölls, S., Tantipolphan, R., Wiggenhorn, M., Winter, G., Jiskoot, W., Friess, W., &Hawe, A. (2012). Particles in therapeutic protein formulations, Part 1: overview of analytical methods. *Journal of pharmaceutical sciences*, *101*(3), 914-935.

CHAPTER 4

PRINCIPLES OF LIQUID CHROMATOGRAPHY IN PROTEOMICS

CONTENTS

4.1 INTRODUCTION

Chromatography is a separation process that separates the elements of a mixture across two states; a free-moving mobile phase and a stable stationary phase. Thin-layer chromatography, paper chromatography, gas chromatography, and liquid chromatography are various types of chromatography, yet they all work on a similar concept (Zhang et al., 2010). The chromatography procedure uses a mixture of molecules that have been dissolved in the solvent. The elements of the mixture may mix with the molecules of both the stationary matrix and solvent when the mobile phase passes over the stationary phase. Due to their varying affinities for every phase, various elements in the mixture flow at various speeds (Cutillas, 2005).

4.2 GENERAL CHROMATOGRAPHIC PRINCIPLES FOR PEPTIDE AND PROTEIN SEGREGATION

Molecules having the minimum affinity for the stationary phase would travel at high speed due to their tendency to retain the solvent, whereas molecules having the maximum affinity would move at slow speed due to their tendency to be linked with the stationary phase and be left behind. This leads to the separation of the mixture into several fractions that may be drawn off and collected separately (Staes et al., 2008).

Due to its flexibility and comparability having mass spectrometry, liquid chromatography (LC) is employed more frequently in proteomics than other chromatography forms. Although gel electrophoresis, LC can separate both peptides and proteins, and may thus be used either before or after 2DGE to pre-fractionate the specimen, segregate peptide mixtures from solitary excised spots, or replace 2DGE as the primary protein segregation method (Van Damme et al., 2009). Various separation principles, including charge, size, hydrophobicity, and affinity for certain ligands, may be used in multiple liquid chromatography procedures. The highest-resolution separations have been accomplished when two or more segregation principles have been applied one after another in orthogonal dimensions, and so is the case with electrophoresis (Lee W & Lee K., 2004).

The stationary phase in proteomics LC techniques is a permeable matrix, generally in the shape of packed beads supported on a certain type of column. The mobile phase which is a solvent having dissolved peptides or proteins travels through the column either naturally or under higher pressure. The rate at which a peptide or protein flows down a column is determined by its affinity for the, and matrices having various physical and chemical characteristics may be utilized to segregate peptides and proteins using various selection principles. The next sections go through these ideas and how they have been used (Staes et al., 2011).

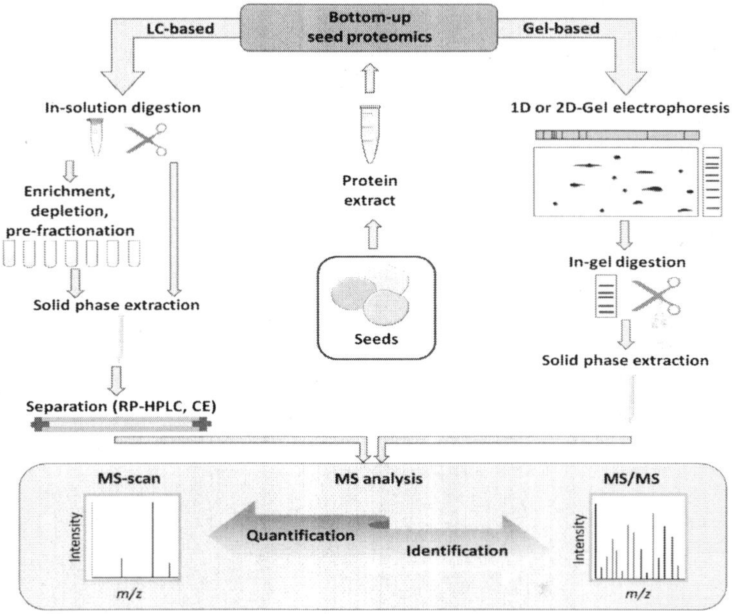

Figure 4.1. A summary of gel-free and gel-based proteomics experimental processes.

Source: Smolikova, Galina & Gorbach, Daria & Lukasheva, Elena & Mavropolo-Stolyarenko, Gregory & Bilova, Tatiana & Soboleva, Alena & Tsarev, Alexander & Romanovskaya, Ekaterina & Podolskaya, Ekaterina & Zhukov, V. & Tikhonovich, Igor & Medvedev, Sergei & Hoehenwarter, Wolfgang & Frolov, Andrej. (2020). Bringing New Methods to the Seed Proteomics Platform: Challenges and Perspectives. International Journal of Molecular Sciences. 21. 9162. 10.3390/ijms21239162.

For peptide and protein separation, multidimensional LC may be employed rather than 2DGE. (b) Without or with a preceding affinity depletion or enhancement step, chromatography processes having alternative separative principles can be utilized as a straight replacement for 2DGE for protein segregation (Gevaert et al., 2006). The second cycle of RP-HPLC is used to feed specific peptide fractions into the mass spectrometer after on-column digestion with trypsin. (c) Complex peptide mixtures can be separated using multidimensional liquid-phase segregation. Because of the complicated structure of the peptide mixture, this method usually invariably includes an affinity enrichment or depletion phase. AC-RPHPLC-CE-MS AC-SEC-RPHPLC-MS and AC-IEC-RPHPLC-MS are three popular methods. The picture depicts mass spectrometry analysis of peptides and proteins (Kaufmann, 1997).

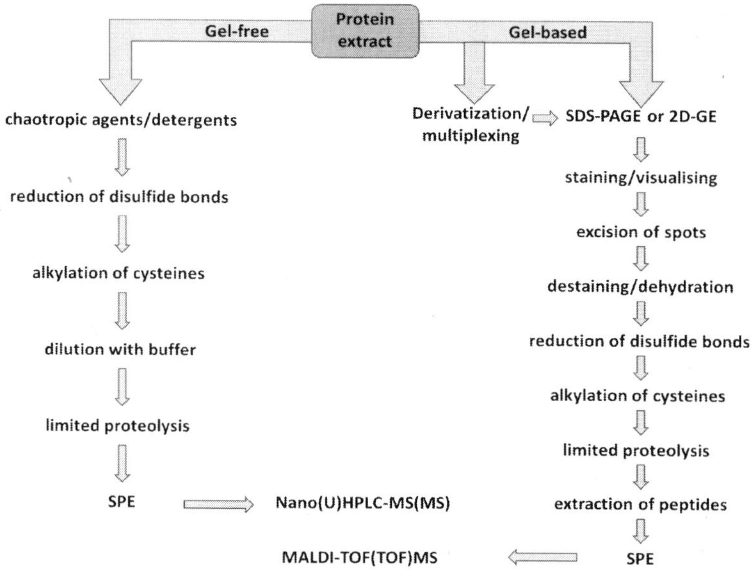

Figure 4.2. Enzymatic digestion is used in both gel-free and gel-based experimental setups.

Source: Smolikova, Galina & Gorbach, Daria & Lukasheva, Elena & Mavropolo-Stolyarenko, Gregory & Bilova, Tatiana & Soboleva, Alena & Tsarev, Alexander & Romanovskaya, Ekaterina & Podolskaya, Ekaterina & Zhukov, V. & Tikhonovich, Igor & Medvedev, Sergei & Hoehenwarter, Wolfgang & Frolov, Andrej. (2020). Bringing New Methods to the Seed Proteomics Platform: Challenges and Perspectives. International Journal of Molecular Sciences. 21. 9162. 10.3390/ijms21239162.

4.3 AFFINITY CHROMATOGRAPHY

Affinity chromatography divides peptides and proteins according to their ligand-binding affinity. An affinity column's matrix comprises ligands that have been extremely selective for specific proteins or protein groups (Walters, 1985). Antibody-coated beads may be used to separate peptides or proteins from a complex mixture, whereas glutathione-coated beads can be used to retain fusion proteins with GST affinity tags. Another kind of affinity chromatography is immobilized metal affinity chromatography (IMAC), which uses positively charged metal ions such as Fe^{3+} or Ga^{3+} in the solid phase. Containing this method, phosphoproteins/peptides, proteins with oligo-histidine affinity tags, and other -ve charged proteins may be distinguished (Hage, 1999).

The first fraction that is typically extracted from the column contains all of the peptides or proteins that had been unable to engage with the affinity matrix. The 2nd fraction, on the other hand, contains all of the peptides or proteins that had been maintained on the column throughout the extraction process. This has been achieved by doing two washes on the affinity matrix. The first solution washed away any proteins that were not connected to the matrix, and the second solution caused any proteins that were attached to the matrix to detach themselves from the matrix (Urh et al., 2009). There are situations in which the initial fraction is necessary, such as when eliminating an excessive amount of protein from a specimen to facilitate an examination of the proteins that are still there. In certain cases, the objective is to segregate the second fraction, which has the proteins which preferentially attach to the affinity matrix. The goal might be to segregate fusion proteins or single, unique proteins with a particular affinity tag. A functional or structural class of proteins (for example, proteins, or that interact with a certain drug or enzyme-substrate) might also be the target. The investigation of phosphor-proteins and other post-translational variations by using affinity chromatography. The separation of proteins that combine to produce a compound is another important use of affinity chromatography (Cuatrecasas& Anfinsen, 1971).

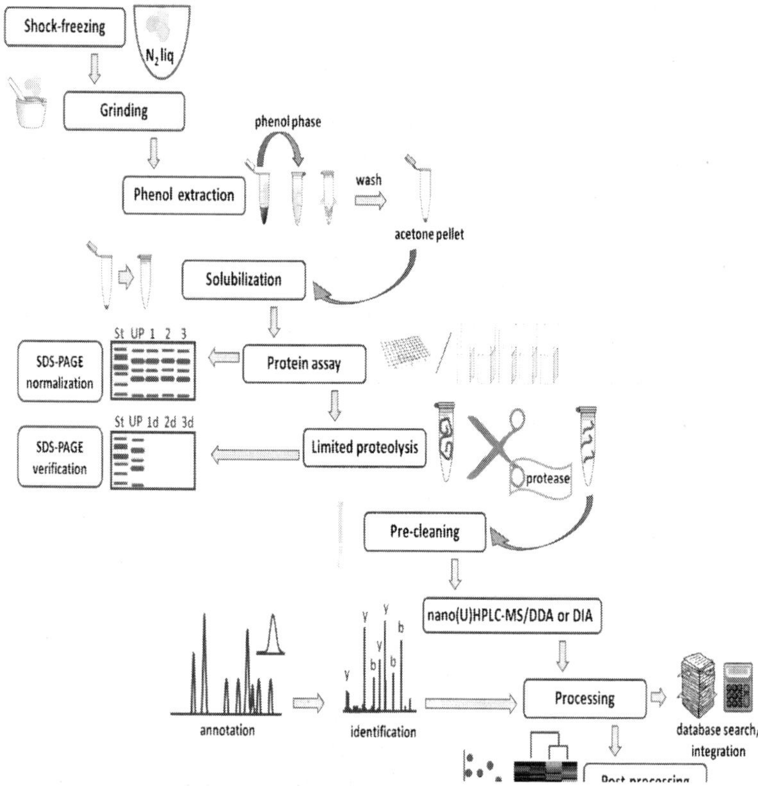

Figure 4.3. Protein separation and specimen preparation are detailed in this LC-based proteomics method.

Source: Smolikova, Galina & Gorbach, Daria & Lukasheva, Elena & Mavropolo-Stolyarenko, Gregory & Bilova, Tatiana & Soboleva, Alena & Tsarev, Alexander & Romanovskaya, Ekaterina & Podolskaya, Ekaterina & Zhukov, V. & Tikhonovich, Igor & Medvedev, Sergei & Hoehenwarter, Wolfgang & Frolov, Andrej. (2020). Bringing New Methods to the Seed Proteomics Platform: Challenges and Perspectives. International Journal of Molecular Sciences. 21. 9162. 10.3390/ijms21239162.

4.4 ION EXCHANGE CHROMATOGRAPHY

Other types of chromatography employed in proteomics, in contrast to affinity chromatography, are nonselective for specific protein classes, that is they have been utilized to profile a specimen instead of targeting individual elements. Ion-exchange chromatography (IEX, IEC) is a technique for separating peptides and proteins based on their charge. It works by reversibly adsorbing solute molecules to a solid phase containing charged chemical groups. Anionic and cationic resins can be utilized to attract molecules in the solvent with opposing charges. Gradient elution is performed by washing the column having buffers of steadily increasing pH or ionic strength, rather than a two-step elution technique (Jungbauer& Hahn, 2009).

Table 4.1. Functional groups utilized on ion exchangers

Anion exchangers	Functional group
Diethylaminoethyl (DEAE)	$-O-CH_2-CH_2-N^+H(CH_2CH_3)_2$
Quaternary aminoethyl (QAE)	$-O-CH_2-CH_2-N^+(C_2H_5)_2-CH_2CHOH-CH_3$
Quaternary ammonium (Q)	$-O-CH_2-CHOH-CH_2-O-CH_2-CHOH-CH_2-N^+(CH_3)_3$
Cation exchangers	Functional group
Carboxymethyl (CM)	$-O-CH_2-COO^-$
Sulfopropyl (SP)	$-O-CH_2-CHOH-CH_2-O-CH_2-CH_2-CH_2SO_3^-$
Methylsulfonate (S)	$-O-CH_2-CHOH-CH_2-O-CH_2-CHOH-CH_2SO_3^-$

A chromatogram is the result of this technique, with the x-axis displaying extraction time and the y-axis displaying absorption maxima that relates to different elements of the specimen. This value is used to define the resolution of a chromatographic separation based on its number of peaks or the maximal capacity that can be discerned from the extraction spectrum's baseline. Comparing peak areas might potentially offer quantitative data, whereas the number of peaks that appear on a chromatogram is reflective of the sample's level of complexity (Fekete et al., 2015).

Chromatofocusing, often known as CF, is a technique that involves making use of an ion-exchange column that has been set to one pH and a buffer that has been calibrated to another pH. The result of this is the formation of a pH gradient through the column, which may be exploited for isoelectric point-based protein extraction. During the operation, focusing effects generate strong peaks and aid in the concentration of distinct fractions (Harinarayan et al., 2006).

4.5 REVERSED-PHASE CHROMATOGRAPHY

Reversed-phase (RP) chromatography, which is similar to ion-exchange chromatography, includes the reversible adsorption of peptides and proteins to the immoveable phase matrix, and RP chromatography produces various fractions through gradient elution. In this particular situation, although, the peptides and proteins are divided as per the given peaks, which correlate to various proteins (A = ovalbumin, B = conalbumin, C = cytochrome c, and D = lysozyme) (Horvath &Melander, 1977).

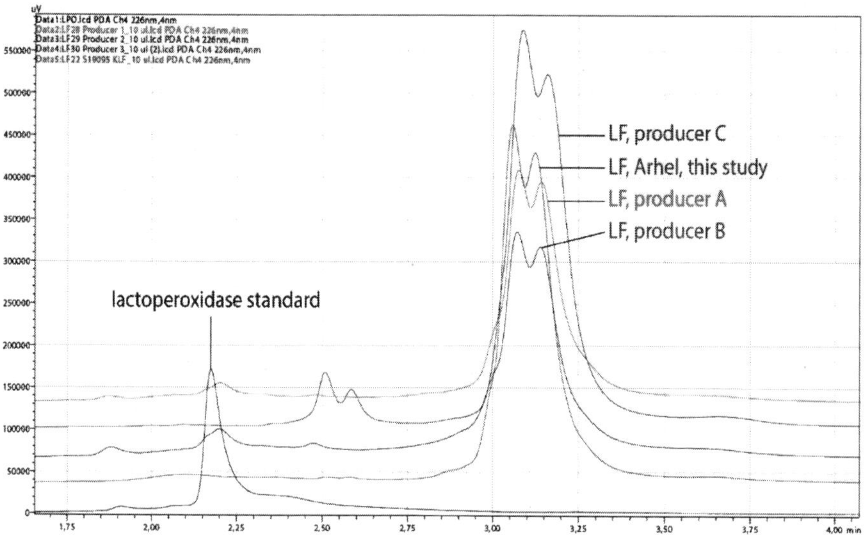

Figure 4.4. Chromatograms displaying the outcomes of protein mixture separations via ion-exchange chromatography.

Source: Matijašić BB, Oberčkal J, Mohar Lorbeg P, Paveljšek D, Skale N, Kolenc B, Gruden Š, Poklar Ulrih N, Kete M, Zupančič Justin M. Characterisation of Lactoferrin Isolated from Acid Whey Using Pilot-Scale Monolithic Ion-Exchange Chromatography. Processes. 2020; 8(7):804. https://doi.org/10.3390/pr8070804

In a particular research article, the first separation (a) had been done at pH 5.85, whereas the second separation (b) had been done at pH 6.5. Operating variables like temperature and pH have a substantial impact on production. Julie Bordonaro and Rebecca Carrier of Rensselaer Polytechnic Institute created the reversed-phase resin, which is made up of hydrophobic ligands like C4 to C18 alkyl groups. The higher performance liquid chromatography (RP-HPLC), wherein the mobile phase is driven through the column at higher pressure, is commonly used in proteomics for reversed-phase separations (Vailaya&Horváth, 1998).

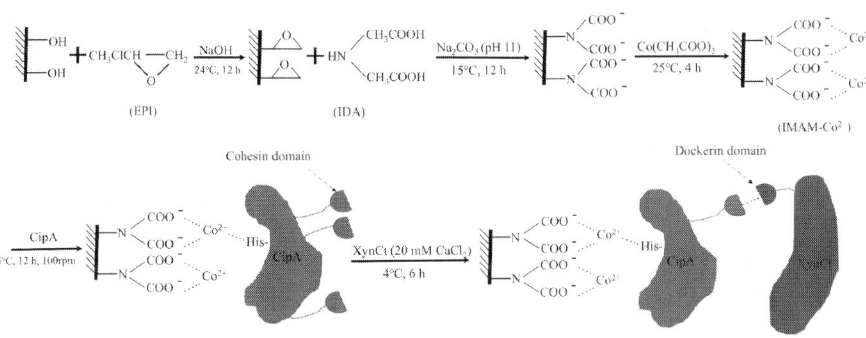

Figure 4.5. On reversed-phase resins, certain frequently utilized n-alkyl hydrocarbon ligands.

Source: Rusli H, Putri RM, Alni A. Recent Developments of Liquid Chromatography Stationary Phases for Compound Separation: From Proteins to Small Organic Compounds. Molecules. 2022; 27(3):907. https://doi.org/10.3390/molecules27030907

Examples include the octyl (C8) ligand, the 2-carbon capping group, and the octadecyl (C18) ligand. RP-HPLC creates quasi-mass-dependent segregation since retention rises as molecular mass increases, but hydrophobicity is the separative principle. To accomplish gradient extraction, first, the weak hydrophobic interactions need to be disrupted, and then the quantity of an organic modifier used in the extraction solution needs to be gradually increased (Dill, 1987). The RP-HPLC chromatographic technique is the most powerful and has the best resolution of all the chromatographic methods that are utilized in proteomics (In practice, a peak capacity reaches up to one hundred components). To enable completely automated peptide segregation and analysis using LC-MS or LC-MS/MS, it is usual practice to connect HPLC columns directly to electrospray ionization mass spectrometers (Dolan, 2002). Proteins are separated using hydrophobic interaction chromatography based on their hydrophobic characteristics. Hydrophobic interaction chromatography, on the other hand, employs polar extraction buffers and a range of resin compositions (aryl ligands or C2–C8 alkyl groups). On either hand, hydrophilic interaction chromatography separates proteins according to their hydrophilic characteristics by employing a polar solid phase in the separation process of protein (Fausnaugh et al., 1984).

4.6 SIZE EXCLUSION CHROMATOGRAPHY

Size exclusion chromatography (sometimes called gel filtration chromatography) has been a profiling method for separating proteins based on size. Inert beads composed of a permeable substance like agarose are put into the column. Smaller proteins may fit through the holes in the beads, therefore they need longer to move along the column than bigger proteins, which cannot fit through the perforations and must go via the spaces between the beads (Barth et al., 1994). The molecular exclusion principle separates the solutes from the stationary phase without requiring a chemical contact. Size exclusion chromatography beads, such as Sepharose, come in a variety of sizes to allow for the separation of peptide or protein mixtures over a wide variety of sizes. (Kostanski et al., 2004)

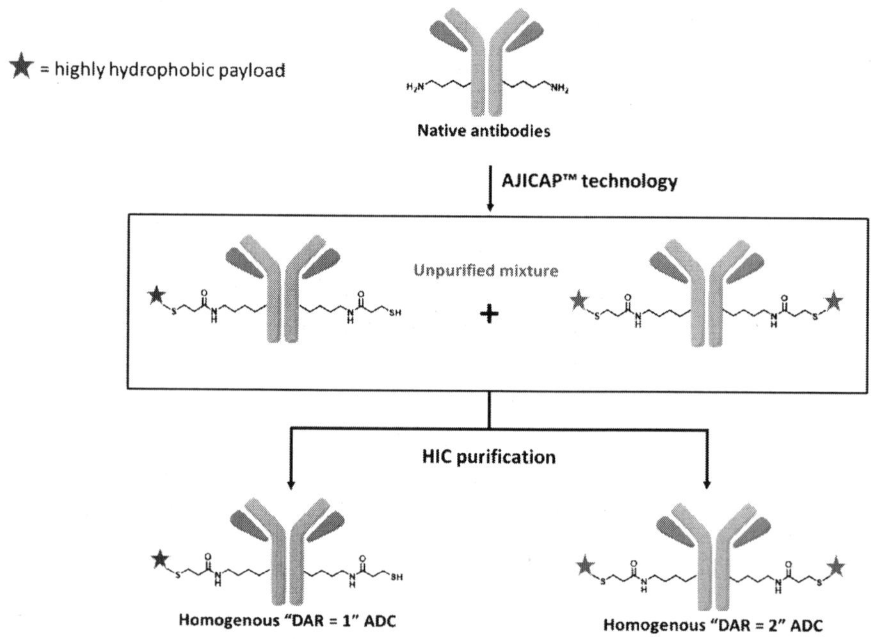

Figure 4.6. Schematic of hydrophobic interaction chromatography that responds to temperature.

Source: Matsuda Y, Leung M, Okuzumi T, Mendelsohn B. A Purification Strategy Utilizing Hydrophobic Interaction Chromatography to Obtain Homogeneous Species from a Site-Specific Antibody Drug Conjugate Produced by AJICAP™ First Generation. Antibodies. 2020; 9(2):16. https://doi.org/10.3390/antib9020016

4.7 MULTIDIMENSIONAL LIQUID CHROMATOGRAPHY

As previously mentioned, liquid chromatography is frequently employed to pre-fractionate materials to segregate tryptic peptides generated from single gel spots, either downstream or upstream of 2DGE (Dugo et al., 2008).

Multidimensional chromatography, on either hand, is an appealing technique to replace 2DGE entirely due to the versatility of Liquid chromatography techniques in terms of mixing multiple separative principles. Liquid chromatography methods get beyond several of the constraints of 2DGE. HPLC columns, for instance, permit high specimen quantities to be loaded and concentrated on the column, making low concentration proteins simpler to identify. Many proteins that have been hard to segregate with 2DGE (for example membrane proteins, and extremely basic proteins) may be readily segregated by utilizing the right resins (Cacciola et al., 2017).

Proteins separated in the liquid state do not need to be stained to be recognized. Because LC columns may be connected directly to the mass spectrometer and because liquid chromatography methods can separate peptides and proteins, the entire analytical procedure, from specimen preparation to peptide mass profiling, can be mechanized (Fairchild et al., 2009). The visual components of protein segregation via 2DGE, such as the pI and molecular mass information that may be obtained from the placements of spots on the gel, have been lost with Liquid chromatography techniques (such information may be utilized in database searches). Several gels having similar specimens may be performed at a similar time with Liquid chromatography, making it a sequential analytic approach. Direct quantitative comparisons between specimens, on either hand, may be made by marking one of the specimens having mass-coded tags that may be recognized in the mass spectrometer (Di Palma et al., 2012).

4.7.1 Multidimensional Liquid Chromatography Techniques in Proteomics

For the investigation of exceedingly complex peptide or protein mixtures, the successive use of multiple chromatographic methods employing distinct physical and chemical separative principles may offer an adequate resolution. For instance, combining ion-exchange chromatography (that segregates proteins based on charge) with RP-HPLC (that segregates proteins based on mass) may reach a similar resolution as 2DGE, with the added benefits of automation, greater sensitivity, and a good example of membrane proteins (Zhang et al., 2010). Furthermore, the practical limits of these multidimensional chromatography methods must be considered. The first thing that has to be looked at is whether or not the solvents and buffers that are utilized in the various stages of each method are appropriate for those stages. The extraction buffer for the 1st-dimension ion exchange step needs to be an appropriate RPHPLC solvent. On the other hand, the elution buffer for the 2nd-dimension RP-HPLC step needs to include the solvents that were employed in the mass spectrometry specimen preparation stage (Wu et al., 2012). A number of the benefits of automation, speed, and resolution will be lost if the fractions had been brought down for cleaning and preparation. However, the solvents and buffers indicated above are suitable, and various researchers have employed ion-exchange accompanied by RP-HPLC-MS to examine the proteomes of species ranging from yeast to humans. HPLC has been almost routinely utilized as the ultimate segregation technique in multidimensional chromatography due to the suitability of RP-HPLC with the solvents utilized in MALDI-MS and ESI-MS (Alternatively, capillary electrophoresis may be employed). Many of the other profiling techniques (ion-exchange chromatography, size exclusion chromatography, and chromate-focusing) are utilized as a first-dimension segregation technique in conjunction with HPLC, often with an affinity chromatography phase preceding it, resulting in a three-dimensional separation approach (Di Palma et al., 2012).

In the beginning, multidimensional chromatography was accomplished through a discontinuous procedure wherein fractions had been first accumulated from the gel filtration or ion exchange column and thereafter manually infused into the HPLC column. This process had been repeated until the desired results had been obtained. The lack of a time limit is one of the benefits of using a discontinuous multidimensional system, even though it requires a lot of manual effort (Okano et al., 2006). It is possible to keep the fractions that elute from the first column off-line constantly, and then feed them one at a time onto the HPLC column that is directly linked to the mass spectrometer. There is also the benefit of being able to apply high specimen volumes to the first column to collect adequate quantities of limited-abundance proteins for examination in the second dimension. This is an additional benefit of the method (Duong et al., 2020).

The first column may be equipped with an automated fraction collection system and a column-switching valve to do away with the requirement of injecting human specimens. After collecting fractions from the first column over the extraction range, the switching valve can bring the RP-HPLC column in line to progressively receive the fractions (Shi et al., 2004). This occurs after the first column has finished eluting its contents. Instead, several researchers have constructed a piece of equipment that comprises a solitary ion exchange system that is connected to several HPLC columns in parallel through the utilization of the appropriate switching valves. Fractions from the ion exchange system have been routed progressively to numerous HPLC columns in this technique, and the cycle is repeated after the first column is regenerated (see Opiteket al.) (Wang &Hanash, 2003).

The use of biphasic columns is the 3rd method for doing multidimensional chromatographic separations. In this method, the proximal section of the column is filled with reversed-phase resin, and the proximal segment contains the ertainc kind of the matrix. This enables the stepwise extraction of fractions from the 1st resin and the gradient extraction of second dimension fractions from the other resin so long as the extraction solvents for the two different kinds of resin do not interact with one another.

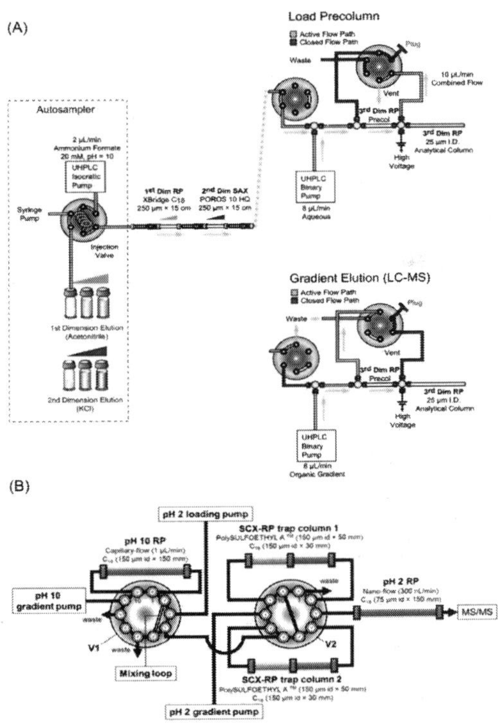

Figure 4.7. Peptide separation using discontinuous multidimensional chromatography.

Source: Duong V-A, Park J-M, Lee H. Review of Three-Dimensional Liquid Chromatography Platforms for Bottom-Up Proteomics. International Journal of Molecular Sciences. *2020; 21(4):1524. https://doi.org/10.3390/ijms21041524*

This technology was initially developed as direct analysis of large protein complexes (DALPC) by colleagues and Link (see Related Reading), and it was later improved into a multidimensional chromatography system by colleagues and Yates (Brown et al., 2020). Proteins from two distinct yeast specimens had been tagged with mass-coded biotin affinity tags which interacted exclusively with cysteine residues throughout this research (that labeling approach is not detailed; it is utilized for comparing protein quantification among specimens). Every protein sample had been digested using trypsin and combined to create a single peptide pool (Wolters et al., 2001). The affinity tag was then selected using affinity chromatography, which reduced the complexity of the mixture by tenfold. Since almost all proteins include a minimum of one cysteine residue, the leftover peptide population covered almost ninety percent of the yeast proteome. Ion exchange chromatography was used to segregate the retrieved peptides utilizing a strong cation exchange resin. Off-line collection of 30 separate specimens resulted in the application of RPHPLC for second-dimension separation on four of the samples (Nägele et al., 2003). *Protein identification technology* (MudPIT) was used. Peptide mixtures were placed onto the ion exchange resin, and first-dimension fractions werereleased into the reversed-phase resin utilizing a stepwise salt gradient. A gradient of acetonitrile was used to extract second-dimension fractions from the reversed-phase resin into the mass spectrometer (Brandão et al., 2019). This procedure, as well as the following regeneration stage, doesn't interact with the ion exchange chromatography phase, and by raising the salt concentration upon regeneration, an additional fraction has been liberated from the ion exchange resin (Karpievitch et al., 2010). Whenever this approach had been implemented to the yeast proteome, it was possible to assign nearly 5000 peptides to 1484 yeast proteins, accounting for almost one-fourth of the yeast proteome. The specimen appeared to cover all protein classes, including those that are often neglected in 2DGE research (Slebos et al., 2008).

REFERENCES

1. Barth, H. G., Jackson, C., &Boyes, B. E. (1994). Size exclusion chromatography. *Analytical Chemistry*, 66(12), 595-620.
2. Brandão, P. F., Duarte, A. C., & Duarte, R. M. (2019). Comprehensive multidimensional liquid chromatography for advancing environmental and natural products research. TrAC Trends in *Analytical Chemistry*, 116, 186-197.
3. Brown, K. A., Tucholski, T., Alpert, A. J., Eken, C., Wesemann, L., Kyrvasilis, A., ... & Ge, Y. (2020). Top-Down Proteomics of Endogenous Membrane Proteins Enabled by Cloud Point Enrichment and Multidimensional Liquid Chromatography–Mass Spectrometry. *Analytical Chemistry*, 92(24), 15726-15735.
4. Cacciola, F., Dugo, P., &Mondello, L. (2017). Multidimensional liquid chromatography in food analysis. *TrAC Trends in Analytical Chemistry*, 96(1), 116-123.
5. Cuatrecasas, P., & Anfinsen, C. B. (1971). Affinity chromatography. *Methods in enzymology*, 22(2), 345-378.
6. Cutillas, P. R. (2005). Principles of nanoflow liquid chromatography and applications to proteomics. *Current Nanoscience*, 1(1), 65-71.
7. Di Palma, S., Hennrich, M. L., Heck, A. J., & Mohammed, S. (2012). Recent advances in peptide separation by multidimensional liquid chromatography for proteome analysis. *Journal of proteomics*, 75(13), 3791-3813.
8. Dill, K. A. (1987). The mechanism of solute retention in reversed-phase liquid chromatography. *Journal of physical chemistry*, 91(7), 1980-1988.
9. Dolan, J. W. (2002). Temperature selectivity in reversed-phase high performance liquid chromatography. *Journal of Chromatography A*, 965(1-2), 195-205.
10. Dugo, P., Cacciola, F., Kumm, T., Dugo, G., &Mondello, L. (2008). Comprehensive multidimensional liquid chromatography: theory and applications. *Journal of Chromatography A*, 1184(1-2), 353-368.
11. Duong, V. A., Park, J. M., & Lee, H. (2020). Review of three-dimensional liquid chromatography platforms for bottom-up proteomics. *International Journal of Molecular Sciences*, 21(4), pp. 1524.

12. Fairchild, J. N., Horváth, K., &Guiochon, G. (2009). Approaches to comprehensive multidimensional liquid chromatography systems. *Journal of Chromatography A*, 1216(9), 1363-1371.
13. Fausnaugh, J. L., Kennedy, L. A., &Regnier, F. E. (1984). Comparison of hydrophobic-interaction and reversed-phase chromatography of proteins. *Journal of Chromatography A*, 317, 141-155.
14. Fekete, S., Beck, A., Veuthey, J. L., &Guillarme, D. (2015). Ion-exchange chromatography for the characterization of biopharmaceuticals. *Journal of pharmaceutical and biomedical analysis*, 113(!), 43-55.
15. Gevaert, K., Van Damme, P., Ghesquière, B., &Vandekerckhove, J. (2006). Protein processing and other modifications analyzed by diagonal peptide chromatography. *Biochimica et Biophysica Acta (BBA)-Proteins and Proteomics*, 1764(12), 1801-1810.
16. Hage, D. S. (1999). Affinity chromatography: a review of clinical applications. *Clinical chemistry*, 45(5), 593-615.
17. Harinarayan, C., Mueller, J., Ljunglöf, A., Fahrner, R., Van Alstine, J., & Van Reis, R. (2006). An exclusion mechanism in ion exchange chromatography. *Biotechnology and Bioengineering*, 95(5), 775-787.
18. Horvath, C., &Melander, W. (1977). Liquid chromatography with hydrocarbonaceous bonded phases; theory and practice of reversed phase chromatography. *Journal of Chromatographic Science*, 15(9), 393-404.
19. Jungbauer, A., & Hahn, R. (2009). Ion-exchange chromatography. *Methods in enzymology*, 463(!), 349-371.
20. Karpievitch, Y. V., Polpitiya, A. D., Anderson, G. A., Smith, R. D., & Dabney, A. R. (2010). Liquid chromatography mass spectrometry-based proteomics: biological and technological aspects. *The annals of applied statistics*, 4(4), 1797.
21. Kaufmann, M. (1997). Unstable proteins: how to subject them to chromatographic separations for purification procedures. *Journal of Chromatography B: Biomedical Sciences and Applications*, 699(1-2), 347-369.
22. Kostanski, L. K., Keller, D. M., &Hamielec, A. E. (2004). Size-exclusion chromatography—a review of calibration methodologies. *Journal of biochemical and biophysical methods*, 58(2), 159-186.
23. Lee, W. C., & Lee, K. H. (2004). Applications of affinity chromatography in proteomics. *Analytical biochemistry*, 324(1), 1-10.

24. Nägele, E., Vollmer, M., &Hörth, P. (2003). Two-dimensional nano-liquid chromatography–mass spectrometry system for applications in proteomics. *Journal of Chromatography A*, 1009(1-2), 197-205.
25. Okano, T., Kondo, T., Kakisaka, T., Fujii, K., Yamada, M., Kato, H., ... &Hirohashi, S. (2006). Plasma proteomics of lung cancer by a linkage of multi-dimensional liquid chromatography and two-dimensional difference gel electrophoresis. *Proteomics*, 6(13), 3938-3948.
26. Shi, Y., Xiang, R., Horváth, C., & Wilkins, J. A. (2004). The role of liquid chromatography in proteomics. *Journal of Chromatography A*, 1053(1-2), 27-36.
27. Slebos, R. J., Brock, J. W., Winters, N. F., Stuart, S. R., Martinez, M. A., Li, M., ... &Liebler, D. C. (2008). Evaluation of strong cation exchange versus isoelectric focusing of peptides for multidimensional liquid chromatography-tandem mass spectrometry. *Journal of proteome research*, 7(12), 5286-5294.
28. Staes, A., Impens, F., Van Damme, P., Ruttens, B., Goethals, M., Demol, H., ... &Gevaert, K. (2011). Selecting protein N-terminal peptides by combined fractional diagonal chromatography. *Nature protocols*, 6(8), 1130-1141.
29. Staes, A., Van Damme, P., Helsens, K., Demol, H., Vandekerckhove, J., &Gevaert, K. (2008). Improved recovery of proteome-informative, protein N-terminal peptides by combined fractional diagonal chromatography (COFRADIC). *Proteomics*, 8(7), 1362-1370.
30. Urh, M., Simpson, D., & Zhao, K. (2009). Affinity chromatography: general methods. *Methods in enzymology*, 463(1), 417-438.
31. Vailaya, A., &Horváth, C. (1998). Retention in reversed-phase chromatography: partition or adsorption? *Journal of Chromatography A*, 829(1-2), 1-27.
32. Van Damme, P., Van Damme, J., Demol, H., Staes, A., Vandekerckhove, J., &Gevaert, K. (2009, December). A review of COFRADIC techniques targeting protein N-terminal acetylation. In BMC proceedings (Vol. 3, No. 6, pp. 1-6). BioMed Central.
33. Walters, R. R. (1985). Affinity chromatography. *Analytical chemistry*, 57(11), 1099A-1114A.
34. Wang, H., &Hanash, S. (2003). Multi-dimensional liquid phase based separations in proteomics. *Journal of Chromatography B*, 787(1), 11-18.

35. Wolters, D. A., Washburn, M. P., & Yates, J. R. (2001). An automated multidimensional protein identification technology for shotgun proteomics. *Analytical chemistry*, 73(23), 5683-5690.
36. Wu, Q., Yuan, H., Zhang, L., & Zhang, Y. (2012). Recent advances on multidimensional liquid chromatography–mass spectrometry for proteomics: From qualitative to quantitative analysis—A review. *Analytica Chimica Acta*, 731, 1-10.
37. Zhang, X., Fang, A., Riley, C. P., Wang, M., Regnier, F. E., & Buck, C. (2010). Multi-dimensional liquid chromatography in proteomics—a review. *Analytica Chimica Acta*, 664(2), 101-113.

CHAPTER 5

PROTEOMICS OF PROTEIN MODIFICATIONS

CONTENTS

5.1 INTRODUCTION

In every organism, there are far more proteins as compared to the genes that code for them. A one-to-one relationship between the quantity of proteins and genes does not occur in higher species; for instance, people have fewer than 25,000 genes however up to 500,000 proteins. The blending of transcripts typically creates numerous messenger RNAs (mRNA) per gene and at least one protein per gene helps in resolving a few of the imbalances in the number of proteins and genes. Other than the splicing process, posttranslational protein alteration is the other primary driver of protein excess in every species. This kind of posttranslational modification happens after or during ribosomes' translation of an mRNA into proteins (Butterfield &Dalle-Donne, 2014).

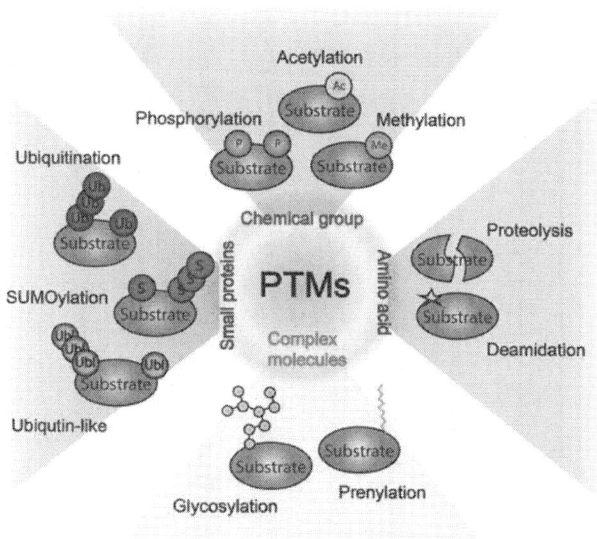

Figure 5.1. Strategies for Post-Translational Modifications (PTMs).

Source: Salas-Lloret D, González-Prieto R. Insights in Post-Translational Modifications: Ubiquitin and SUMO. International Journal of Molecular Sciences. 2022; 23(6):3281. https://doi.org/10.3390/ijms23063281

Proteins need posttranslational change for their unique function and their storage, disintegration, and regulation of many biochemical functions. Some proteins, for instance, should be phosphorylated, which requires the addition of one or even more PO4 groups to the protein backbone (Arrell et al., 2001).

Proteins remain functional in the signaling pathways, division of the cell, as well as other processes in an entity once they have been phosphorylated. 1/3rd of all proteins in eukaryotic organisms are phosphorylated (Repetto et al., 2003).

Similarly, glycosylation includes the insertion of carbohydrate monomers into 50% of all proteins in humans. Binding to fatty acids or lipids modifies some membrane proteins. The formation of disulfide connections between distinct components stabilizes several other proteins, such as insulin and immunoglobulin (Soufi ET AL., 2012). Several additional proteins are degraded before becoming active; for instance, insulin is translated as a preproinsulin from mRNA. It is destroyed by removing terminal amino acids to create proinsulin, which is then further destroyed to produce insulin, which acts as a hormone to govern glucose absorption in mammalian cells. Furthermore, some proteins are ubiquitinated by the inclusion of a protein family known as ubiquitin, which marks these proteins for breakdown by the protease system (Rudolf et al., 2013). Many additional proteins, like arginine to citrulline, asparagine to aspartic acid, or glutamine to glutamic acid, may be modified by changing one amino acid to the other. Some proteins may have a stretch of amino acids termed "intein" removed from their core at times. Intein elimination is aided by a protein splicing mechanism analogous to RNA splicing in the synthesis of transcribed. Phosphorylation, glycosylation, and ubiquitination are the three most common changes that proteins go through. These alterations were previously identified by studying one protein at a time, and now that proteomic technologies employing mass spectrometry are available, these changes may be easily demonstrated in a huge range of proteins. This section covers this and many other posttranscriptional alterations that alter the pattern or composition of amino acids in proteins (Desiderio&Nibbering, 2006).

5.2 PHOSPHORYLATION AND PHOSPHOPROTEOMICS

This post-transcriptional alteration exists in above 30% of mammalian proteins. This entails adding one or maybe more PO4 molecules to certain amino acids in a protein. PO4 groups are attached to the protein amino acids threonine, serine, and tyrosine in human cells. In comparison, prokaryotes phosphorylate amino acids like aspartic, glutamic, and histidine, while eukaryotic phosphorylate tyrosine, serine, and threonine (Jers et al., 2008; Lemeer& Heck, 2009). Sometimes, phosphorylation happens in arginine, lysine, and cysteine residues of proteins in either prokaryotes or eukaryotes. In mammalian cells, the proportion of phosphorylation of the three amino acid residues is 1000:100:1 for threonine, serine, as well as tyrosine, respectively. Proteins are phosphorylated at several sites, and a cell typically contains a variety of phosphorylated isomers with varying amounts of phosphorylation. The introduction of the PO4 group via phosphorylation imparts a negative charge to the proteins (Ajadi et al., 2020).

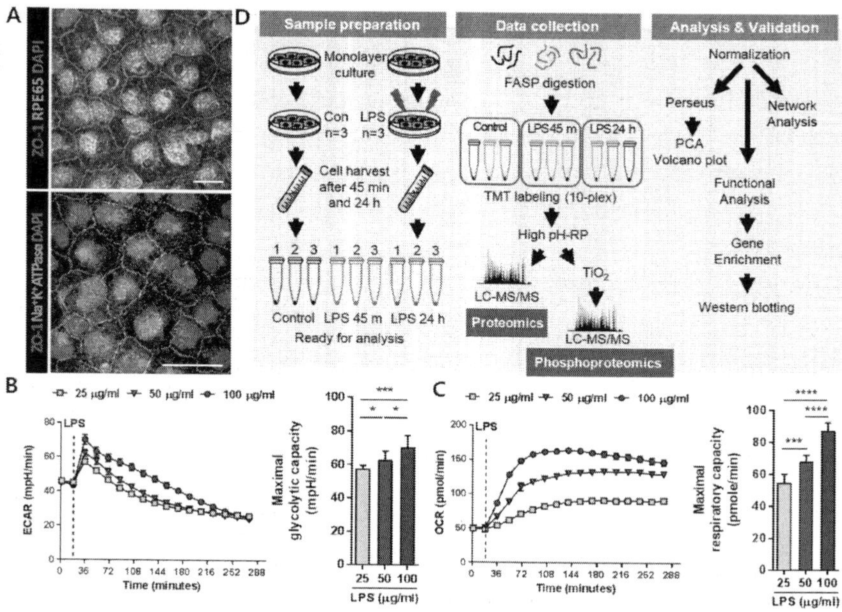

Figure 5.2. Protein Phosphorylation as well as Phosphoproteome.

Source: Song J, Han D, Lee H, Kim DJ, Cho J-Y, Park J-H, Seok SH. A Comprehensive Proteomic and Phosphoproteomic Analysis of Retinal Pigment Epithelium Reveals Multiple Pathway Alterations in Response to the Inflammatory Stimuli. International Journal of Molecular Sciences. 2020; 21(9):3037. https://doi.org/10.3390/ijms21093037

Protein kinases are responsible for phosphorylation. Dephosphorylation is the process of removing the phosphate group from phosphorylated proteins (s). Protein phosphatase catalyzes the dephosphorylation process. About 2% of the human genome codes for protein kinases and protein phosphatases. In humans, about 500 protein kinases and over 100 protein phosphatases have been identified (Nakagami et al., 2010). There are roughly 120 kinase genes and 40 phosphatase genes in yeast. Typically, phosphorylation affects the activation or inactivation of a protein and its biological or metabolic function. Following dephosphorylation, a phosphorylated protein that is functional might become inert. Similarly, a phosphorylated protein that is inert turns active following dephosphorylation. Proteins are phosphorylated to activate specific mechanisms, like signal transmission and cell division, and cancer-causing paths in mammalian cells (Engholm□Keller & Larsen, 2013). The effective treatment of some types of human cancer with Gleevec (Novartis Pharmaceuticals Corp., East Hanover, NJ), a blocker of BCR-Abl protein kinase, demonstrates the importance of phosphorylation in cancer. Protein kinases are regarded as a good targeted therapy for cancer as well as other disorders caused by signal transduction pathway disruption. A concerted attempt is required to discover blockers of various protein kinases to cure cancer and other disorders (Nühse et al., 2004).

Phosphoproteomics is the extensive investigation of genes associated with the phosphorylation process. A huge number of enzymes and the various phosphorylation sites are analyzed in a single effort. In comparison to past methods, which detailed the biochemistry and genetics of a single protein at a period, this method describes the biochemistry and genetics of several proteins simultaneously (Pinkse et al., 2008). Phosphoproteomics is now feasible as a result of technological advancements such as mass spectrometry in the study of proteins as well as the existence of gene and protein databanksPhosphoproteins and their locations of phosphorylation are recognized by the rise in molecular mass of the peptides acquired after mass spectrometric comparative method of their amino acid or inferred nucleotide pattern with the pattern data given in the protein as well as gene databanks (Umezawa et al., 2013).

Phosphoproteomics requires many processes, including the production of protein samples, their concentration, and tryptic digestion before mass spectrometric detection (Olsen et al., 2010).

Typically, the test proteins are produced using a variety of procedures. Between these techniques, phosphoprotein isotope-coded affinity tag (PhIAT), isotope-coded affinity tag (ICAT), as well as carbon isotope tagging with amino acids in cell culture are used often (SILAC). PhIAT directly inserts isotopes into protein phosphoserine as well as phosphothreonine residues (Hardman et al., 2019).

Besides phosphorylation sites, the various two techniques incorporate isotopes into the protein at other locations. *In vitro* proteins are labeled with PhIAT and ICAT, while *in vivo* proteins are labeled with SILAC. SILAC is beneficial for *in vivo* tagging of proteins in cell cultures maintained under circumstances that may affect the degree of phosphorylation. In Chapter 3, the ICAT and SILAC methodologies are explained (Van Bentem et al., 2006).

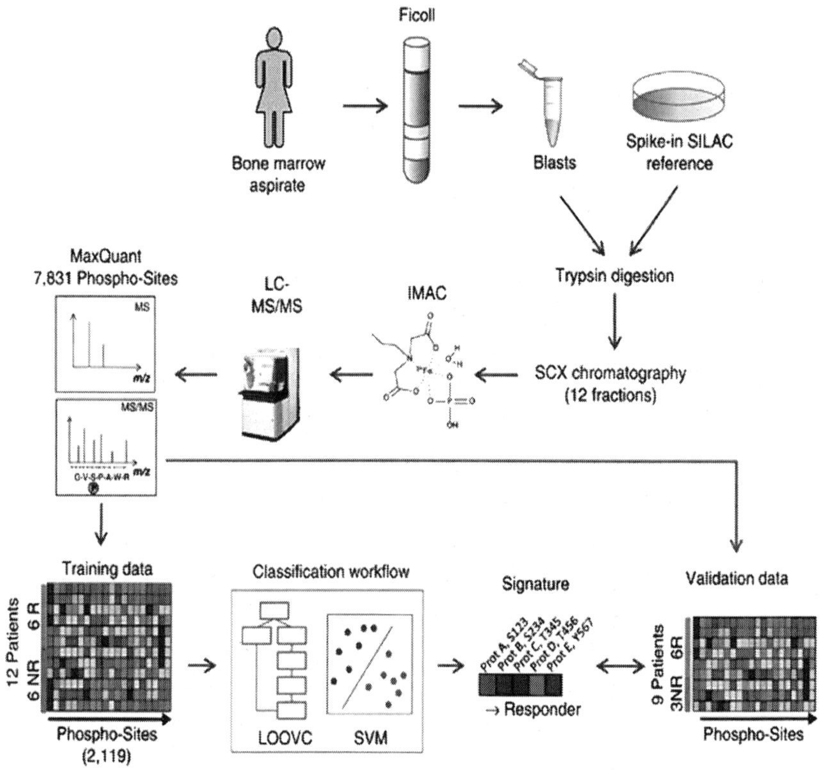

Figure 5.3. Global phosphoproteome analysis of human bone marrow reveals predictive phosphorylation markers.

Source: Schaab, C., Oppermann, F., Klammer, M. et al. Global phosphoproteome analysis of human bone marrow reveals predictive phosphorylation markers for the treatment of acute myeloid leukemia with quizartinib. Leukemia 28, 716–719 (2014). https://doi.org/10.1038/leu.2013.347

On the affinities, column phosphoproteins are enhanced. SILAC is often used in tandem with immobilized metal affinity chromatography (IMAC). The column including the anti-phosphotyrosine antibody is utilized to concentrate phosphotyrosine-containing protein due to their poor interaction with phosphoproteins, phosphoserine and phosphothreonine antibodies could not be utilized to concentrate phosphoproteins on a column (Lin et al., 2009). Nevertheless, they are employed to detect phosphoproteins on the gel. These antibodies are all widely accessible. As explained in Chapter 3, the optimized phosphoproteins are digested with trypsin; phosphorylated and unphosphorylated specimens are then combined and evaluated using a mass spectrometer. Following mass spectrometry, phosphoproteins are distinguished by changes in molecular weights depending on the phosphate group connected to the amino acids of peptides. Proteins are confirmed by analyzing their amino acid patterns to those in the protein database (Schmidt et al., 2014).

Utilizing this approach for the investigation of phosphoproteins, a great number of novel phosphoproteins and phosphorylation locations in many paths have been found. Using mass spectrometry, the many elements of the signaling pathways and other control mechanisms have been discovered (Roux &Thibault, 2013). For instance, phosphoproteomics has improved our understanding of yeast mating pheromone-induced signaling cascades. And over 500 proteins and 729 phosphorylation sites in the yeast signal transduction pathway are receptive to the mating pheromone, and 139 of these phosphorylation sites were changed in response to the pheromone in yeast. In addition, this work sheds light on the function of phosphorylation in the RNA-processing and transport paths that regulate mRNA metabolism in yeast (Mithoe&Menke, 2011). The phosphoproteomic technique has established the impact of phosphorylation on the subunits of the epidermal growth factor (EGF) receptor as well as their involvement in mRNA metabolism in mammals. These investigations have also improved our understanding of the signal transduction of extracellular regulated kinase (ERK) protein kinase. Numerous new phosphorylation sites have been found in tumor suppressor proteins (TSC1 and TSC2). By identifying the elements of such a system, phosphoproteomics has thrown considerable light on the function of phosphorylation in the signaling route. Phosphorylation research is very relevant since phosphorylation sites are consistent between organisms (Wong et al., 2019).

5.3 GLYCOSYLATION AND GLYCOSYLATION

5.3.1 Glycosylation

Protein glycosylation is a common posttranslational alteration. It includes integrating a glucose moiety into proteins. Glycoproteins are proteins that have been glycosylated. Glycoproteins make over half of all proteins in mammalian cells. They regulate the durability, anchoring, cell-to-cell connections, antigenic and immunological specialization, secretion, reproduction, pathogenesis protection, and practically all other activities of proteins in the cell, as outlined in the following part (Hart, 1992).

Carbohydrates and proteins are the two primary elements of glycosylated proteins. The carbohydrate portion may range anywhere from 2% to 80% of the overall bulk. Proteoglycans are glycoproteins that have a large percentage of carbohydrates, as compared to glycoproteins that have a modest fraction of carbs. Glycoproteins, on the other hand, refer to all glycosylated proteins (Haltiwanger et al., 2004).

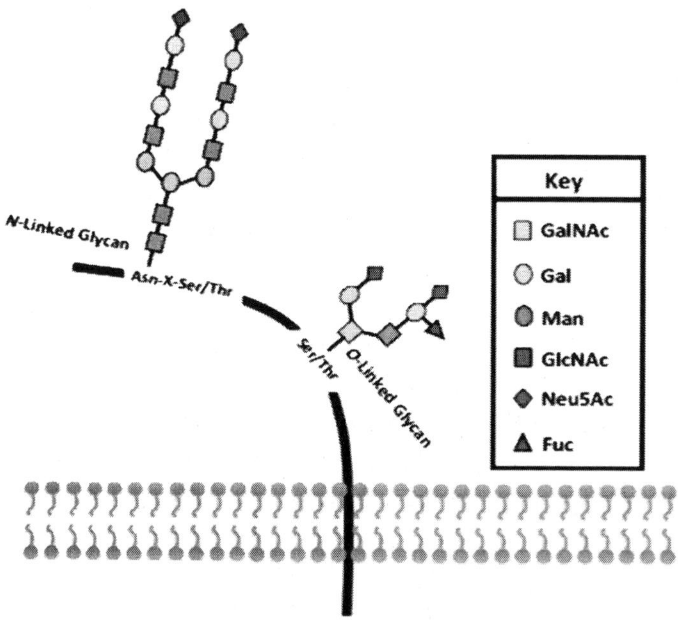

Figure 5.4. The two primary kinds of protein glycosylation are shown in Figure 5.4. Attaching sugar moieties to protein is a post-translational alteration that gives the proteins more proteome variety.

Source: Nardy, Ana & Freire de Lima, Leonardo & Freire-de-Lima, Celio & Morrot, Alexandre. (2016). The Sweet Side of Immune Evasion: Role of Glycans in the Mechanisms of Cancer Progression. Frontiers in Oncology. 6. 10.3389/fonc.2016.00054.

Carbohydrate moieties are chemically attached to asparagine, serine, and threonine in glycoproteins. Carbohydrate connections may be divided into two types. Carbohydrates are connected to the aromatic ring of asparagines for one type of protein known as "N-linked glycoproteins." Carbohydrates are attached to the OH group of serine or threonine in the next type of coupling, resulting in "O-linked glycoproteins" (Rudd et al., 2001).

Acetylglucosamine is coupled to the N atom of the NH2 group in the asparagine of N-linked glycoproteins. Galactose or glucosyl-galactose is bonded to the O atom of serine or threonine in O-linked glycoproteins. Additional sugar molecules, like arabinose or mannose, may bind to serine, threonine, or proline in O-linked glycoproteins at sometimes. In O-linked glycoproteins, many different carbohydrates may be connected to proteins (Lis & Sharon, 1993). A glycoprotein may have just one kind of sugar molecule linkage or both O-linked and N-linked sugar molecules. Glycophorin, which has 15 O-linked sugar molecules but one N-linked sugar molecule, is an instance of a glycoprotein with both links. Glycophorin is found in the erythrocyte cell membrane. The protein and carbohydrate constituents of glycoprotein dictate its characteristics. The polymeric variants of the glycophorin gene, which codes for distinct amino acids at the 5 and 26 locations in the protein part of this glycoprotein, define the M, N, and MN blood types in people (Ohtsubo &Marth, 2006). One of the immediate impacts of glycosylation is that it enhances the solubility of proteins by making them more hydrophilic due to the existence of numerous OH groups on the carbohydrate elements. Glycoproteins are required for practically all cell processes. Glycoproteins have a variety of activities, including cell adhesion, antigenic, antibody, enzymatic, as well as hormonal functions; they also perform a part in fertility and cytoskeleton (Reily et al., 2019).

Glycoproteins help a cell adhere to some other cell and regulate cell-to-cell contacts. N-CAM is a glycoprotein that lets nerve cells detect and attach. This glycoprotein also helps nerve cells and muscle cells connect at the neuromuscular junction. Fibroblasts would attach to every ligand or molecule that contains fibronectin. Antigens and antibodies are both glycoproteins. The existence of glycoprotein in the cell surface of the blood cell causes the antigenic feature of the human blood group (Kahne et al., 1989). Antigen A is found in blood group A, antigen B is found in blood group B, antigen A and B are found in blood group AB, and antigen A and B are not found in blood group O. N-acetyl galactosamine glycosylates the protein A in antigen. Galactose glycosylates the protein B in antigen. By eliminating their sugar moiety or substituting one sugar on the blood cell membrane with another, they might lose or modify their antigenic properties (Pinho& Reis, 2015). Antibody immunoglobulins are glycoproteins that lose their ability to act as antibodies when the sugar component is removed. Glycoproteins are a kind of enzyme that includes oxidoreductases, hydrolases, and transferases; the latter can also act as a blocker of some enzymes. Glycoproteins include hormones such as human chorionic gonadotropin (hCG), which is detected in human female urine throughout pregnancy, and erythropoietin, which regulates erythrocyte synthesis. Glycoproteins are also used to transport hormones, vitamins, and cations. Many glycoproteins are involved in the human reproductive process (Varki et al., 2009). Glycoproteins improve sperm attractiveness to the egg, enable sperm entry to the cervix, regulate sperm penetrance to the zona pellucida, and inhibit polyspermy, or egg fertilization by more than one sperm. Glycoproteins play an important role in cell structure. Glycoproteins are found in cartilage, synaptosomes, axons, and microsomes, among other places (Stowell et al., 2015). Blood clotting proteins including thrombin, prothrombin, and fibrinogen are also glycoproteins. The capsules present in S cell bacteria contain glycoproteins that coat the outside of the cell wall, giving the S cells a smooth look. Glycoproteins are also found in bacterial flagella, which govern bacterial motility. Furthermore, glycoproteins like mucin play many additional roles, including cell protection, internal organ protection, and skin protection (Steen et al., 1998).

5.3.2 Glycoproteomics

Due to their many functions in the construction and functioning of cells in various animals, glycoproteins are very important for human health. Numerous altered glycoproteins serve as disease indicators in humans. Certain glycoproteins are modified in several malignancies. The prototypical case is prostate-specific antigen (PSA) (Wuhrer et al., 2007). Glycoproteins act as diagnostic markers for many different malignancies, such as breast cancer. They also offer a foundation for the immunotherapeutic and pharmacological treatment of human disorders. Glycoproteins are excellent pharmacological and drug combination sites. Herceptin treatment is a great illustration of immunotherapy for specific tumors (Polasky et al., 2020).

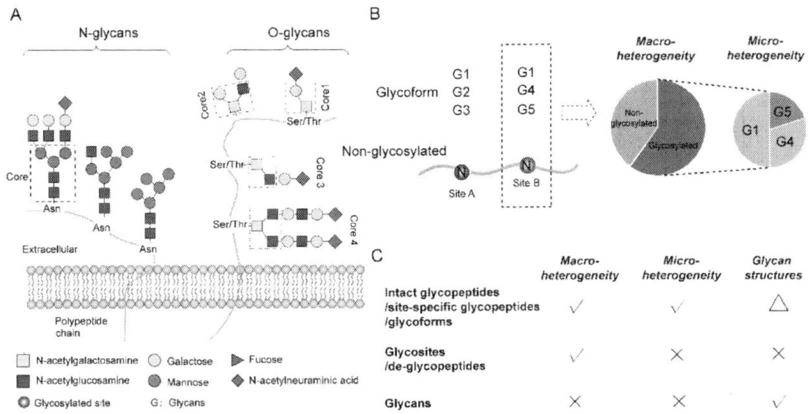

Figure 5.5. Glycans are ubiquitous. High-throughput glycoproteomics techniques provide information.

Source; Fang P, Ji Y, Oellerich T, Urlaub H, Pan K-T. Strategies for Proteome-Wide Quantification of Glycosylation Macro- and Micro-Heterogeneity. International Journal of Molecular Sciences. 2022; 23(3):1609. https://doi.org/10.3390/ijms23031609

Consequently, knowing the nature of various glycoproteins has garnered considerable attention. Various technologies have been implemented to examine their structure, such as the type of the proteins, their glycosylation, as well as the binding site of certain carbohydrate monomers or polymers (Tissot et al., 2009). These approaches use affinity chromatography to enrich glycoproteins and mass spectrometry to determine the type of proteins and conjugated carbohydrates. These approaches are rapid and high-throughput, and they offer data on a wide number of glycoproteins in a little amount of time. Some insoluble membrane glycoproteins are turned accessible by trypsin hydrolysis, and the resultant soluble glycopeptides are concentrated by chromatographic techniques and evaluated by mass spectrometry. The enriching of glycoproteins may be accomplished using two distinct but significant techniques. The first technique employs a family of plant proteins known as "lectin." Lectins link only to a certain class of glycoproteins (Tian & Zhang, 2010). Consequently, glycoproteins may be readily isolated using affinity chromatography over a column that identifies a specific lectin. Lectins have been utilized to see and compare different glycoproteins on microchips. Therefore, lectins cannot be used to affinity-purify all glycoproteins. Thus, different approaches for the enhancement of glycoproteins have been proposed. Just one of those techniques, which was first established by Aebersold et al. (2003), is built on hydrazine chemistry. Initially, glycoproteins are degraded by contact with periodate, and they'll be attached to hydrazide resin in a column. The column is rinsed many times to remove unbound glycoproteins from the hydrazide column. Eventually, the attached glycoproteins are freed from the matrix by enzymatic hydrolysis that hydrolyzes the link for both the glycerol molecule of glycoproteins as well as the NH2 position of the hydrazide in the resin (Palaniappan&Bertozzi, 2016).

5.4 UBIQUITINATION AND UBIQUITINOMICS

5.4.1 Ubiquitination or Ubiquitinylation

Ubiquitin is a tiny, evolutionarily conserved protein that is only present in eukaryotes. This protein's properties were determined in the 1980s. Aaron Ciechanover, Avram Hersko, and Irwin Rose received the Nobel Prize in Chemistry in 2004 for their great accomplishment (Welchman et al., 2005).

Ubiquitination is the method of assigning ubiquitin to a protein, marking it for enzymatic destruction by proteosomes. Because all proteins are destroyed at the time of destruction, ubiquitination is a nearly ubiquitous alteration of proteins. The half-life of proteins varies; most have a half-life of several seconds, while others contain a half-life of many weeks. The half-life of hemoglobin is nearly three weeks. It's still unclear what factors influence a protein's lifetime (Randles& Walters, 2012). However it is not clear why the presence of a specific amino acid at the N-terminal would govern a protein's service life, the N-terminal rule predicts that proteins with aspartic acid at the N-terminal are relatively brief and proteins with serine at the N-terminal are sustained over time. The ubiquitination procedure is often used to assess a protein's longevity (Hicke et al., 2005). In addition to maintaining protein stability, ubiquitin also plays a role in the cell stage, DNA repair, and translation. Ubiquitin is a 76-amino-acid protein that is tiny yet is involved in gene regulation. Just three amino acid changes occur across the ubiquitin protein in yeast as well as in humans, indicating that it is conserved. The lysine residue of ubiquitin is bound to the glycine residue at the C-terminus of a protein, which would be subsequently designated for destruction by the proteasome (Abbas et al., 2008). Many additional ubiquitin molecules may be included in the ubiquitin molecule after it has been attached to the protein. Monoubiquitination occurs when only one ubiquitin molecule is connected to the protein of interest (Yao et al., 2011), polyubiquitination occurs while a few ubiquitin molecules are attached in tandem to the first ubiquitin, and multiubiquitination occurs when multiple ubiquitin molecules are connected to the protein of interest at distinct lysine residues. Protein trafficking is aided by monoubiquitination, but protein breakdown is aided by polyubiquitination. Additional to protein stability, the diverse types of ubiquitinated molecules may affect other activities of the targeting proteins, although these functions remain unknown (Hoeller&Dikic, 2009).

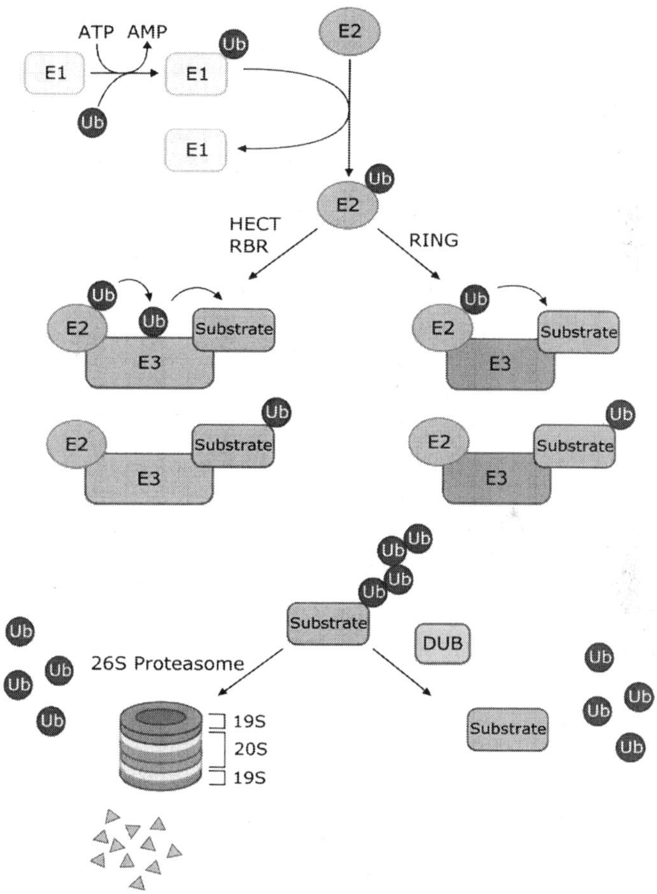

Figure 5.6. Brief-Introduction-to-Ubiquitin-and-Protein-Ubiquitination.

Source; LaPlante G, Zhang W. Targeting the Ubiquitin-Proteasome System for Cancer Therapeutics by Small-Molecule Inhibitors. Cancers. 2021; 13(12):3079. https://doi.org/10.3390/cancers13123079

The ubiquitination mechanism involves at least three different types of enzymes. The El activating enzyme, E2 conjugating enzyme, and E3 ligase are among them. El activates ubiquitin in a resource way by converting adenosine triphosphate (ATP) to adenosine diphosphate (ADP). E2 and E3 bind ubiquitin to the protein that has been designated for destruction (Weissman et al., 2011). The ubiquitinated protein subsequently binds to the 26S proteasome through a collector or the 19S regulatory component of the proteasome straight. The proteosome's 20S catalytic component degrades the designated protein. Besides these types of enzymes, another group of enzymes is engaged in the elimination of ubiquitin from proteosomes. The deubiquitinating enzymes are a class of enzymes (DUB). There are 500–600 ligases and roughly 70 DUB enzymes in people. The ubiquitination mechanism and the involvement of several enzymes inside this mechanism are thoroughly explained (Rolfe et al., 1995).

5.4.2 Proteomics of Ubiquitin Modifications

Ubiquitin changes proteins that are damaged, mistranslated, or biologically changed to have become dysfunctional; these proteins are then designated to be eliminated by proteasome hydrolysis. The proteomics of ubiquitin alterations was initially investigated in yeast (Peng et al., 2003). These researchers replicated the ubiquitin gene & inserted a 6x histidine tag just at the start of the gene. The cloned ubiquitin genes with histidine labels were transferred into yeast cells bearing a loss in the yeast chromosomes' native ubiquitin gene (Swatek&Komander, 2016). These cells synthesized ubiquitin proteins using histidine hexamers. A nickel column was used to purify the ubiquitin linked with conjugating proteins and the proteins intended for elimination. Mass spectrometry was used to evaluate the peptides generated following tryptic digestion of ubiquitin-attached proteins, and they were recognized by a match with the protein pattern in the protein databank. Utilizing such an innovative technique, Peng et al. (2003) discovered over 1000 ubiquitination sites in 72 yeast ubiquitin-protein conjugates.

The researchers also discovered alterations in seven ubiquitin lysine residues in protein conjugates, suggesting that polyubiquitin in yeast is diversifying in vivo. This strategy developed by Peng et al. (2003) serves as the foundation for the same method in other creatures, such as humans. Following their technique of cloned ubiquitin genes with histidine tags were transfected into human hepatocytes, and several ubiquitin-associated proteins, including the conjugating and DUB enzymes, were identified in human cells. Furthermore, proteasome inhibitors were employed to extract as well as to characterize ubiquitin-associated proteins in humans following mass spectrometry investigation (Kim et al., 2011). An examination of the ubiquitination process throughout different organisms, such as humans, mice, worms, flies, and yeast, found that the proportion of E2, E3, and DUB enzymes is proportional to the complexity of an organism. In mice, many novel enzymes have been identified, including 4 E1, 13 E2, 97 E3, and 6 DUB. Errors in the ubiquitination mechanism have been linked to illnesses like Alzheimer's, Parkinson's, autoimmune, and cancer (Peng et al., 2003).

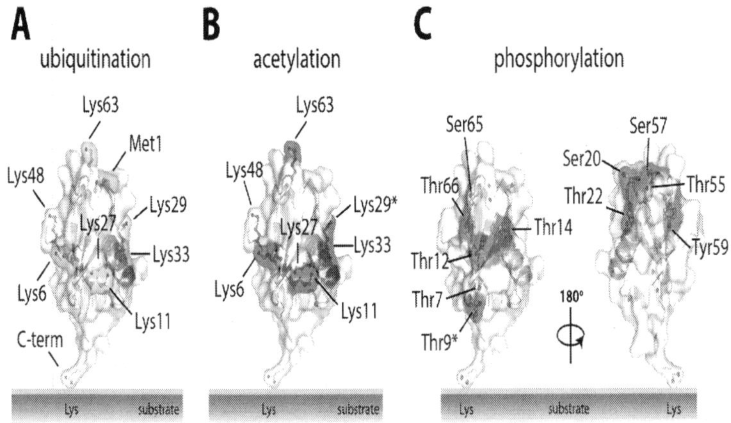

Figure 5.7. Modification sites on ubiquitin.

Source: Swatek, K., Komander, D. Ubiquitin modifications. Cell Res 26, 399–422 (2016). https://doi.org/10.1038/cr.2016.39

5.5 MISCELLANEOUS MODIFICATIONS OF PROTEINS

Other than phosphorylation, glycosylation, and ubiquitination, proteins result in a variety of additional changes. Proteolysis, acetylation, methylation, sulfonation, frensylation, and sumoylation are a few examples (Barraud &Tisné, 2019).

5.5.1 Proteolysis

The extent of the chain of amino acids in proteins is reduced via proteolysis. The N-terminal amino acid methionine, which serves as the starting amino acid in all proteins, is eliminated as quickly as and before creatine supplementation is complete. Additionally, some proteins are first generated as a longer chain, which is then proteolytically cleaved to produce an even smaller active version of the enzyme and protein. Insulin and zymogen are common instances of this kind of protein. Such proteins are generated as pre-pro-proteins and afterward through proteolysis to give initially pro-proteins and subsequently proteins, that are their active component (King et al., 1996).

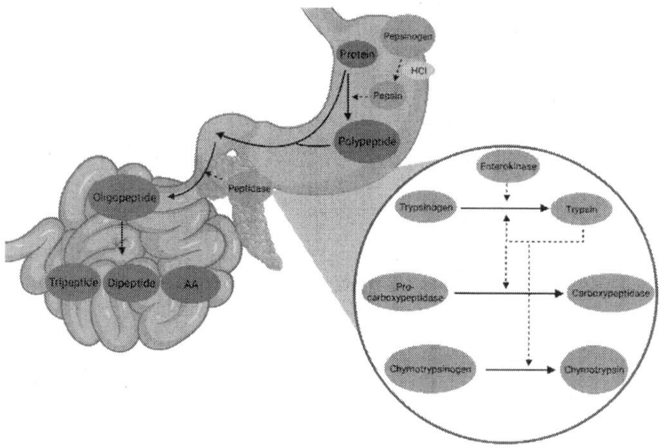

Figure 5.8. Proteolysis is the breakdown of proteins into smaller polypeptides or amino acids.

Source; Kurz A, Seifert J. Factors Influencing Proteolysis and Protein Utilization in the Intestine of Pigs: A Review. Animals. 2021; 11(12):3551. https://doi.org/10.3390/ani11123551

5.5.2 Methylation

The introduction of a methyl or CH_3 group to the lysine residue in many proteins causes methylation. Histone methylation is essential for gene activity regulation (Costello &Plass, 2001).

5.5.3 Sulfation

Certain proteins, such as gastrin, are sulfated by the attachment of a sulfate group or SO_4 to a tyrosine receptor. Sulfation is accomplished by two enzymatic reactions controlled by two distinct transferases (Waqif et al., 1997).

5.5.4 Prenylation

Some proteins, like Ras and transducin, have isoprenoid groups linked to their cysteine residues. Farnesyl and geranyl are isoprenoids with 15 and 20 carbons, correspondingly (Zhang, & Casey, 1996).

5.5.5 Hydroxylation and Carboxylation

The insertion of the OH group to proline plus lysine sites in the protein results in hydroxylation. Hydroxylation occurs in the synthesis of vitamin C, which acts as a complement for the corresponding hydroxylases. Some proteins can carboxylate glutamine residues. Vitamin K is required as a component in this mechanism (Chakraborty & Coates, (2005).

REFERENCES

1. Abbas, T., Sivaprasad, U., Terai, K., Amador, V., Pagano, M., & Dutta, A. (2008). PCNA-dependent regulation of p21 ubiquitylation and degradation via the CRL4Cdt2 ubiquitin ligase complex. *Genes & development*, 22(18), 2496-2506.
2. Ajadi, A. A., Cisse, A., Ahmad, S., Yifeng, W., Yazhou, S. H. U., Shufan, L. I., ... & Jian, Z. (2020). Protein phosphorylation and phosphoproteome: An overview of rice. *Rice Science*, 27(3), 184-200.
3. Arrell, D. K., Neverova, I., & Van Eyk, J. E. (2001). Cardiovascular proteomics: evolution and potential. *Circulation Research*, 88(8), 763-773.
4. Barraud, P., &Tisné, C. (2019). To be or not to be modified: miscellaneous aspects influencing nucleotide modifications in tRNAs. IUBMB life, 71(8), 1126-1140.
5. Butterfield, D. A., &Dalle-Donne, I. (2014). Redox proteomics: from protein modifications to cellular dysfunction and disease. *Mass spectrometry reviews*, 33(1), 1-6.
6. Chakraborty, R., & Coates, J. D. (2005). Hydroxylation and carboxylation—two crucial steps of anaerobic benzene degradation by Dechloromonas strain RCB. *Applied and Environmental Microbiology*, 71(9), 5427-5432.
7. Costello, J. F., &Plass, C. (2001). Methylation matters. *Journal of medical genetics*, 38(5), 285-303.
8. Desiderio, D. M., &Nibbering, N. M. (2006). Redox proteomics: from protein modifications to cellular dysfunction and diseases (Vol. 9, pp. 5). John Wiley & Sons.
9. Engholm-Keller, K., & Larsen, M. R. (2013). Technologies and challenges in large-scale phosphoproteomics. *Proteomics*, 13(6), 910-931.
10. Haltiwanger, R. S., & Lowe, J. B. (2004). Role of glycosylation in development. *Annual review of biochemistry*, 73(1), 491-537.
11. Hardman, G., Perkins, S., Brownridge, P. J., Clarke, C. J., Byrne, D. P., Campbell, A. E., ... &Eyers, C. E. (2019). Strong anion exchange-mediated phosphoproteomics reveals extensive human non-canonical phosphorylation. *The EMBO journal*, 38(21), e100847.
12. Hart, G. W. (1992). Glycosylation. *Current opinion in cell biology*, 4(6), 1017-1023.

13. Hicke, L., Schubert, H. L., & Hill, C. P. (2005). Ubiquitin-binding domains. *Nature reviews Molecular cell biology*, 6(8), 610-621.
14. Hoeller, D., &Dikic, I. (2009). Targeting the ubiquitin system in cancer therapy. *Nature*, 458(7237), 438-444.
15. Jers, C., Soufi, B., Grangeasse, C., Deutscher, J., &Mijakovic, I. (2008). Phosphoproteomics in bacteria: towards a systemic understanding of bacterial phosphorylation networks. *Expert review of proteomics*, 5(4), 619-627.
16. Kahne, D., Walker, S., Cheng, Y., & Van Engen, D. (1989). Glycosylation of unreactive substrates. *Journal of the American Chemical Society*, 111(17), 6881-6882.
17. Kim, W., Bennett, E. J., Huttlin, E. L., Guo, A., Li, J., Possemato, A., ... &Gygi, S. P. (2011). Systematic and quantitative assessment of the ubiquitin-modified proteome. *Molecular cell*, 44(2), 325-340.
18. King, R. W., Deshaies, R. J., Peters, J. M., &Kirschner, M. W. (1996). How proteolysis drives the cell cycle. *Science*, 274(5293), 1652-1659.
19. Lemeer, S., & Heck, A. J. (2009). The phosphoproteomics data explosion. *Current opinion in chemical biology*, 13(4), 414-420.
20. Lin, M. H., Hsu, T. L., Lin, S. Y., Pan, Y. J., Jan, J. T., Wang, J. T., ... & Wu, S. H. (2009). Phosphoproteomics of *Klebsiella pneumoniae* NTUH-K2044 reveals a tight link between tyrosine phosphorylation and virulence. *Molecular & Cellular Proteomics*, 8(12), 2613-2623.
21. Lis, H., & Sharon, N. (1993). Protein glycosylation: structural and functional aspects. *European journal of biochemistry*, 218(1), 1-27.
22. Mithoe, S. C., &Menke, F. L. (2011). Phosphoproteomics perspective on plant signal transduction and tyrosine phosphorylation. *Phytochemistry*, 72(10), 997-1006.
23. Nakagami, H., Sugiyama, N., Mochida, K., Daudi, A., Yoshida, Y., Toyoda, T., ... &Shirasu, K. (2010). Large-scale comparative phosphoproteomics identifies conserved phosphorylation sites in plants. *Plant Physiology*, 153(3), 1161-1174.
24. Nühse, T. S., Stensballe, A., Jensen, O. N., & Peck, S. C. (2004). Phosphoproteomics of the Arabidopsis plasma membrane and a new phosphorylation site database. *The Plant Cell*, 16(9), 2394-2405.
25. Ohtsubo, K., &Marth, J. D. (2006). Glycosylation in cellular mechanisms of health and disease. *Cell*, 126(5), 855-867.

26. Olsen, J. V., Vermeulen, M., Santamaria, A., Kumar, C., Miller, M. L., Jensen, L. J., ... & Mann, M. (2010). Quantitative phosphoproteomics reveals widespread full phosphorylation site occupancy during mitosis. *Science signaling*, 3(104), ra3-ra3.
27. Palaniappan, K. K., &Bertozzi, C. R. (2016). Chemical glycoproteomics. *Chemical reviews*, 116(23), 14277-14306.
28. Peng, J., Schwartz, D., Elias, J. E., Thoreen, C. C., Cheng, D., Marsischky, G., ... &Gygi, S. P. (2003). A proteomics approach to understanding protein ubiquitination. *Nature Biotechnology*, 21(8), 921-926.
29. Pinho, S. S., & Reis, C. A. (2015). Glycosylation in cancer: mechanisms and clinical implications. *Nature Reviews Cancer*, 15(9), 540-555.
30. Pinkse, M. W., Mohammed, S., Gouw, J. W., van Breukelen, B., Vos, H. R., & Heck, A. J. (2008). Highly robust, automated, and sensitive online TiO2-based phosphoproteomics applied to study endogenous phosphorylation in *Drosophila melanogaster*. *Journal of proteome research*, 7(2), 687-697.
31. Polasky, D. A., Yu, F., Teo, G. C., &Nesvizhskii, A. I. (2020). Fast and comprehensive N-and O-glycoproteomics analysis with MSFragger-Glyco. *Nature methods*, 17(11), 1125-1132.
32. Randles, L., & Walters, K. J. (2012). Ubiquitin and its binding domains. *Frontiers in bioscience* (Landmark edition), 17, 2140.
33. Reily, C., Stewart, T. J., Renfrow, M. B., & Novak, J. (2019). Glycosylation in health and disease. *Nature Reviews Nephrology*, 15(6), 346-366.
34. Repetto, O., Bestel-Corre, G., Dumas-Gaudot, E., Berta, G., Gianinazzi-Pearson, V., &Gianinazzi, S. (2003). Targeted proteomics to identify cadmium-induced protein modifications in *Glomus mosseae*-inoculated pea roots. *New Phytologist*, 157(3), 555-567.
35. Rolfe, M., Beer-Romero, P., Glass, S., Eckstein, J., Berdo, I., Theodoras, A., ... &Draetta, G. (1995). Reconstitution of p53-ubiquitinylation reactions from purified components: the role of human ubiquitin-conjugating enzyme UBC4 and E6-associated protein (E6AP). *Proceedings of the National Academy of Sciences*, 92(8), 3264-3268.
36. Roux, P. P., &Thibault, P. (2013). The coming of age of phosphoproteomics—from large data sets to inference of protein functions. *Molecular & Cellular Proteomics*, 12(12), 3453-3464.

37. Rudd, P. M., Elliott, T., Cresswell, P., Wilson, I. A., &Dwek, R. A. (2001). Glycosylation and the immune system. *Science*, 291(5512), 2370-2376.
38. Rudolf, G. C., Heydenreuter, W., &Sieber, S. A. (2013). Chemical proteomics: ligation and cleavage of protein modifications. *Current Opinion in Chemical Biology*, 17(1), 110-117.
39. Schmidt, A., Trentini, D. B., Spiess, S., Fuhrmann, J., Ammerer, G., Mechtler, K., & Clausen, T. (2014). Quantitative phosphoproteomics reveals the role of protein arginine phosphorylation in the bacterial stress response. *Molecular & Cellular Proteomics*, 13(2), 537-550.
40. Soufi, B., Soares, N. C., Ravikumar, V., &Macek, B. (2012). Proteomics reveals evidence of cross-talk between protein modifications in bacteria: focus on acetylation and phosphorylation. *Current opinion in microbiology*, 15(3), 357-363.
41. Steen, P. V. D., Rudd, P. M., Dwek, R. A., &Opdenakker, G. (1998). Concepts and principles of O-linked glycosylation. *Critical reviews in biochemistry and molecular biology*, 33(3), 151-208.
42. Stowell, S. R., Ju, T., & Cummings, R. D. (2015). Protein glycosylation in cancer. Annual Review of Pathology: *Mechanisms of Disease*, 10, 473-510.
43. Swatek, K. N., &Komander, D. (2016). Ubiquitin modifications. *Cell research*, 26(4), 399-422.
44. Tian, Y., & Zhang, H. (2010). Glycoproteomics and clinical applications. *Proteomics–Clinical Applications*, 4(2), 124-132.
45. Tissot, B., North, S. J., Ceroni, A., Pang, P. C., Panico, M., Rosati, F., ... & Morris, H. R. (2009). Glycoproteomics: past, present and future. *FEBS letters*, 583(11), 1728-1735.
46. Umezawa, T., Sugiyama, N., Takahashi, F., Anderson, J. C., Ishihama, Y., Peck, S. C., & Shinozaki, K. (2013). Genetics and phosphoproteomics reveal a protein phosphorylation network in the abscisic acid signaling pathway in *Arabidopsis thaliana. Science signaling*, 6(270), rs8-rs8.
47. Van Bentem, S. D. L. F., Anrather, D., Roitinger, E., Djamei, A., Hufnagl, T., Barta, A., ... &Hirt, H. (2006). Phosphoproteomics reveals extensive *in vivo* phosphorylation of Arabidopsis proteins involved in RNA metabolism. *Nucleic acids research*, 34(11), 3267-3278.
48. Varki, A., Kannagi, R., & Toole, B. P. (2009). Glycosylation changes in cancer. Essentials of Glycobiology. 2nd edition, (pp. 1-5).

49. Waqif, M., Bazin, P., Saur, O., Lavalley, J. C., Blanchard, G., &Touret, O. (1997). Study of ceria sulfation. *Applied Catalysis B: Environmental*, 11(2), 193-205.

50. Weissman, A. M., Shabek, N., &Ciechanover, A. (2011). The predator becomes the prey: regulating the ubiquitin system by ubiquitylation and degradation. *Nature reviews Molecular cell biology*, 12(9), 605-620.

51. Welchman, R. L., Gordon, C., & Mayer, R. J. (2005). Ubiquitin and ubiquitin-like proteins as multifunctional signals. *Nature reviews Molecular cell biology*, 6(8), 599-609.

52. Wong, M. M., Bhaskara, G. B., Wen, T. N., Lin, W. D., Nguyen, T. T., Chong, G. L., &Verslues, P. E. (2019). Phosphoproteomics of Arabidopsis Highly ABA-Induced1 identifies AT-Hook–Like10 phosphorylation required for stress growth regulation. *Proceedings of the National Academy of Sciences,* 116(6), 2354-2363.

53. Wuhrer, M., Catalina, M. I., Deelder, A. M., &Hokke, C. H. (2007). Glycoproteomics based on tandem mass spectrometry of glycopeptides. *Journal of Chromatography B*, 849(1-2), 115-128.

54. Yao, Q., Li, H., Liu, B. Q., Huang, X. Y., &Guo, L. (2011). SUMOylation-regulated protein phosphorylation, evidence from quantitative phosphoproteomics analyses. *Journal of Biological Chemistry*, 286(31), 27342-27349.

55. Zhang, F. L., & Casey, P. J. (1996). Protein prenylation: molecular mechanisms and functional consequences. *Annual review of biochemistry*, 65(1), 241-269.

CHAPTER 6

PROTEOMICS OF PROTEIN-PROTEIN INTERACTOMES

CONTENTS

6.1 INTRODUCTION

The one-gene–one-enzyme hypothesis led to the idea that biological processes are catalyzed individually by multiple enzymes in a biochemical way after the other. This viewpoint was reinforced by the advancement of molecular biology (Cafarelli et al., 2017). This concept of one enzyme performing one biochemical process in separation has been replaced by modern knowledge that a type of protein connects in a metabolic path, thanks to the advent of systems biology. This new technique led to the discovery of protein connections and the idea of interactomes. Since interactomes include several proteins involved in various metabolic processes, they are also known as complexosomes. Interactome analysis is becoming a critical element in improving protein complexes and activity (Chua & Wong, 2008).

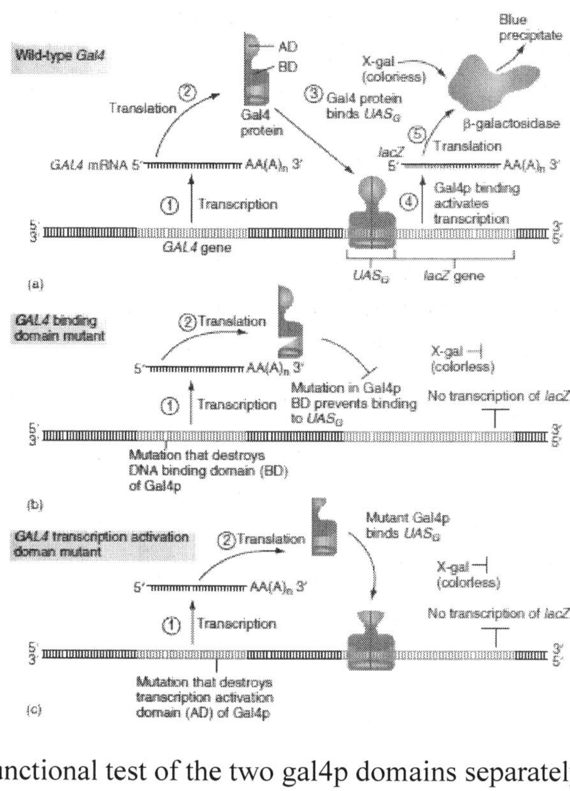

Figure 6.1. Functional test of the two gal4p domains separately.

Source: McLaughlin, W.A., Chen, K., Hou, T. et al. On the detection of functionally coherent groups of protein domains with an extension to protein annotation. BMC Bioinformatics 8, 390 (2007). https://doi.org/10.1186/1471-2105-8-390

For a comprehensive knowledge of all metabolic pathways, a thorough comprehension of interactomes is required. Furthermore, a better knowledge of interactomes is crucial to understanding illnesses and drug development, though either disease and drug discovery are linked to alterations in metabolic processes regulated by multiple proteins cooperating in interactomes. In yeast, almost full knowledge of interactomes has been achieved. Techniques for studying protein interactomes *in vivo* and *in vitro* have been created. This section presents a few of these ways (Rual et al., 2005).

6.2 PROTEIN-PROTEIN INTERACTIONS *IN VIVO*

6.2.1 Yeast Two-Hybrid Assay for Protein-Protein Interactions

The majority of proteins have many motifs or domains with specialized functions. Typically, a protein is known as a Transcription activator (TA) or transcription factor that facilitates DNA transcription (TF) (Causier& Davies, 2002). At a minimum, two domains comprise the transcription factor: The one which connects with the DNA (to be transcribed) at a particular place, termed the DNA–binding domain (BD), and another that binds with RNA polymerase to activate transcription, termed the Activation Domain (AD). Those regions of the TA/TF protein must be encoded by a DNA portion. In terms of bringing those two domains together to allow the transcription of a gene, they have to be capable of binding whether on a similar DNA segment or distinct segments. Yeast GAL-4 is a transcription activator, and the yeast two-hybrid method makes it possible to indicate the existence of the BD and AD by activating a beta-galactosidase reporter gene. The transcription of the beta-galactosidase reporter gene in the yeast-two hybrid bacterium is deduced from its ability to produce blue yeast colonies in the addition of a chromogenic material in the growing medium (Walhout& Vidal, 2001).

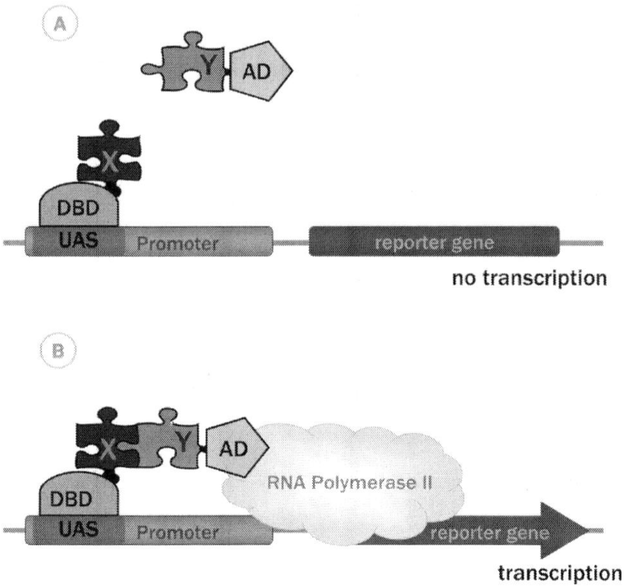

Figure 6.2. The yeast two-hybrid system involves two Yep plasmids.

Source: Brückner A, Polge C, Lentze N, Auerbach D, Schlattner U. Yeast Two-Hybrid, a Powerful Tool for Systems Biology. International Journal of Molecular Sciences. 2009; 10(6):2763-2788. https://doi.org/10.3390/ijms10062763

Fields and Song conceptualized the yeast two-hybrid (Y2H) technology (1989). By expressing a reporter gene, the association of many proteins with a given protein is determined (Fields & Song, 1989). This approach is predicated on the notion as well that the DNA binding domain (BD) and the activation domain (AD) are necessary for the transcription of a gene. This utilizes the yeast gal-4 gene system. The gal-4 gene makes a protein which is a lac gene transcription factor. The lac gene provides instructions for making the beta-galactosidase enzyme in the existence of a gal-4 gene product containing a DNA BD and an AD. This enzyme catalyzes the transformation of a colorless chemical into a blue product that may be seen in developing yeast colonies on a plate (Luo et al., 1997). Therefore, the activation of the lac gene signals that the gal-4 gene is operating properly due to the contact between the two zones. If some of such domains are dysfunctional, the lac gene is not translated, and yeast colonies look translucent in the lack of beta-galactosidase. As soon as these two disciplines stay in close vicinity, although if they occur on two distinct peptides, they may trigger the transcription of the beta-galactosidase gene in yeast, resulting in the blue coloration of yeast colonies. Therefore, this system offers a mechanism for determining the contact between different proteins comprising the gal-4 DNA binding domain and the transcriptional activation domain. If such two proteins combine to put these two gal-4 regions into close contact, the yeast colonies look blue. If such proteins do not bind, then there is no translation of the galactosidase gene and the yeast colonies stay colorless (Sato et al., 1994). The gal-4 DNA segment encoding the BD is fused with the DNA segment encoding the protein involved in the protein-protein interaction. Similarly, the gal-4 AD DNA is fused with some other DNA fragment expressing the protein whose relationship with the protein expressed by the DNA fragment encoding the BD is being investigated. Such two chimeric gene complexes including the BD and AD are utilized to transplant yeast cells. If such two proteins combine to put the BD and AD into close contact, transfected yeast cells will form blue colonies. In the addition of a chromogenic material, the network connections engage the lac gene and create beta-galactosidase, which alters the color of yeast colonies (Lentze&Auerbach, 2008).

In their initial two-hybrid system, Fields and Song (1989) fused DNA expressing 2 main associating proteins (SNF1 and SNF4) to target DNA encoding gal-4 BD and gal-4 AD, correspondingly. The yeast cells were transfected using these plasmid variants. The interaction between the chimeric proteins created in the yeast cells brought along the BD with AD of the gal-4 transcription factor, resulting in the transcriptional activation of the beta-galactosidase gene as well as the formation of blue colonies in the developing yeast cells. The construct holding BD was referred to as "bait" in this system, whereas the construct holding AD was referred to as "prey." The chimeric protein having a peptide other than SNF4 or the other appropriate peptide users can interact with AD was not susceptible to bait having the SNF1 and BD sequences (Bartel et al., 1993). For their purpose of restoring the transcriptional activation of the beta-galactosidase gene and generating blue colonies in increasing yeast cells transfected with suitable structures in pairwise pairings, this system proved to be a valuable tool for identifying the conversing proteins with AD vs. a recognized bait comprising BD (Serebriiskii et al., 2000).

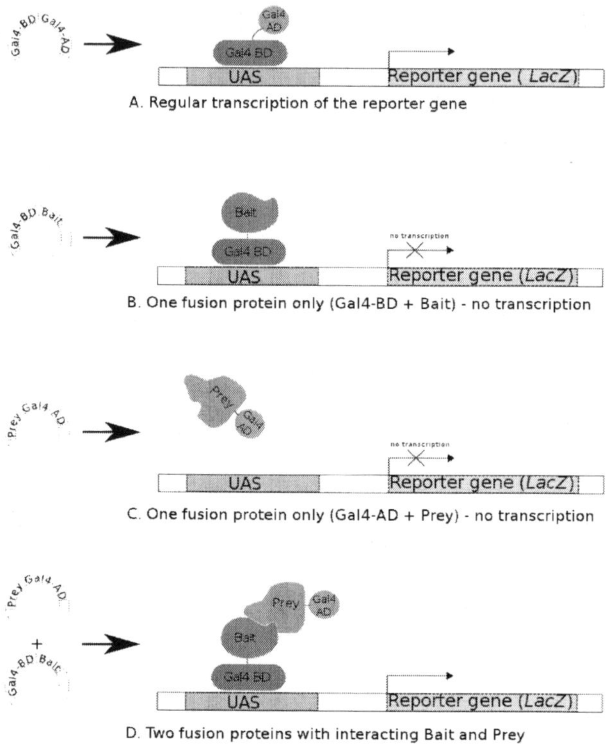

Figure 6.3. Graphical illustration of the yeast two-hybrid system.

Source: By Anna - Own work, CC BY-SA 3.0, https://commons.wikimedia.org/w/index.php?curid=2890233

Because since conception, many improvements have been made to the yeast two-hybrid system, including the utilization of a reporting scheme apart from galactosidase. Additionally, the yeast two-hybrid system may be made to function by fusing the alpha mating kind cell further with a different mating-type cell, i.e., a yeast cell independently transfected with Plasmid DNA having ORF for protein complexes including BD and AD (Fields, 2005). The development of the yeast two-hybrid system by mating is seen as favorable since it provides for the posttranslational changes of the proteins necessary for their associations. Additionally, it has been conceivable to develop a hybrid system to investigate protein interactions utilizing *Escherichia coli* or even mammalian cells as opposed to yeast cells. In yeast and other species, the yeast two-hybrid method has been crucial in building the web of protein complexes and interactomes. Figures above illustrate the function of several yeast two-hybrid system components, including the BD and AD (James et al., 1996).

6.2.2 Phage Display

Since the associated protein is shown or generated on the exterior of the bacterial virus, this process is termed "phage display." In this approach, the gene producing the desired protein is cloned alongside the target DNA producing the phage coat protein (Smith &Petrenko, 1997). The bacteria then are infected with it. Upon infecting bacteria with the reconstructed phage genome, new disease particles are produced that produce the desired protein on the coat protein's exterior. These viral proteins are selected by engaging with an antibody vs. a protein on the well's membrane; the virus particles connect directly to the antibody-containing well while others are swept away. Such virus particles are utilized to attack a different microbial host to create only offspring virions with the protein of interest on the viral coat's surface. Isolated proteins out from coat protein are examined by mass spectrometry to determine which protein matches the sequence of amino acids in the protein database (Azzazy& Highsmith, 2002).

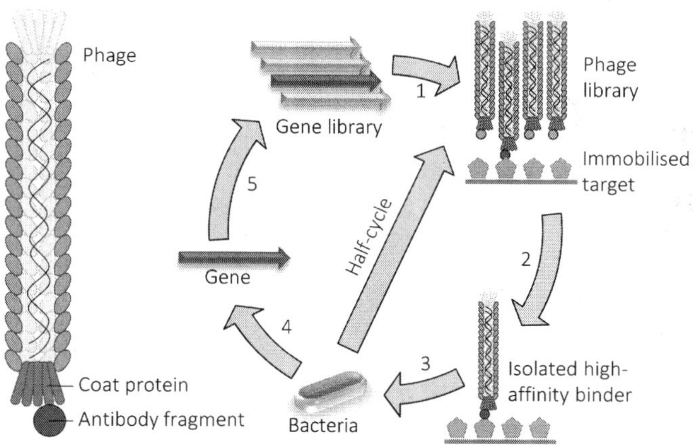

Figure 6.4. Visualization of Phage Display.

Source:https://en.wikipedia.org/wiki/File:Phage_display.png

6.3 ANALYSIS OF PROTEIN INTERACTION *IN VITRO*

In vitro, protein-protein contacts may be identified in a variety of methods. Coimmunoprecipitation, protein pulldown assay, chemical crosslinking, fluorescence resonance energy transfer (FRET), label transfer, and tandem affinity purification (TAP) are these methods. TAP is the first technique with high throughput (Piehler, 2005).

The conventional approach for demonstrating the existence of protein complexes is to coimmunoprecipitate proteins with an antibody. Whenever a protein is caused by a primary antibody, the inter-acting protein(s) are also prompted together with the protein usually generated by the antigen-antibody contact; upon electrophoresis, the elements of the associated proteins are viewed by Western analysis within the same band on the gel (Galarneau et al., 2002).

Protein pulldown is similar to coimmunoprecipitation, with the exception that rather than using an antibody, a ligand is used to identify interacting proteins on gel electrophoresis. The interacting proteins' combined movement on the gel seems to be much shorter as compared to the protein molecules in the interacting group. By transferring a radioactive tag from one protein to the other, a tag transmission can be utilized to find the sample with an insufficient or transient interplay (Doi et al., 2002).

TAP entails cleaning of associated proteins depending on their affinity for another molecule confined to a matrix, followed by mass spectrometric identifying associated proteins. Although this is a high-throughput technique, it is unable to identify weak transitory interactionssince such proteins are isolated with the affinity purification (Kawahashi et al., 2003). In several organisms, such as yeast, the TAP technique has been utilized to define interactomes or protein systems. Following is a short explanation of this technique.

6.3.1 TAP and Mass Spectrometry

Proteins of interest are extracted using an epitope label in this approach. The epitope tags are short stretches of amino acids fused to the proteins of relevance. Purification of proteins with so epitopes is accomplished by column chromatography, which primarily involves a protein extract over a matrix containing a protein having an affinity for the epitope, like Sepharose beads (Sigma Aldrich Co., St. Louis, MO) (Montiel et al., 2018). The proteins are kept on the matrix surface and bind to the affinity protein-coupled to Sepharose beads throughout this kind of column chromatography. Two epitopes are likely connected to the protein sample. The TEV protease site separates the immunoglobin G (IgG) binding domain (protein A) and the calmodulin-binding domain (CBD).

TEV-Protease site
N— — —,—,— — —,— — — —C
Epitope #1 Epitope #2 Protein #1
(ProtA) (CBD)

A protein mixture is initially run over an IgG-containing Sepharose column in this technique, and the bound proteins are then extracted by changing the ionic strength of the elution buffer. Following that, the proteins are digested using TEV proteases to eliminate the initial epitope with IgG high affinity. The protein mixture is then run through a second Sepharose column containing calmodulin linked to the beads. The CBD-containing proteins are kept in this column and are acquired by changing the ionic strength of the elution buffer (Goeury et al., 2019).

Mass spectrometry is then used to identify such the proteins of relevance and their interactions. Some proteins, particularly ones responsible for DNA reactions (replication, repair, recombination, and transcription), may be isolated straight on a DNA Sepharose column with no need for epitope tags since they easily bind to the DNA connected to the Sepharose beads in the matrix. A variation in the elution buffer's ionic content removes the proteins attached to the matrix. Mass spectrometry is used to identify the proteins retrieved (Takaku et al., 1995).

6.3.2 Mass Spectrometric Identification of Interacting Proteins

The proteins are prepared when TAP is put into a spectrometer, and the various proteins are recognized. In such a study, stable isotopes are tagged on a specimen of protein affinity isolated to use the epitope tag and matched to the abundances of distinct peaks in the two samples to discover the associating proteins in a given composition. As a result, the quantitative ICAT mass spectrometry presented in the chapter is substantially the same (Jauregui et al., 1997). The proteins are separated with no epitope tag on a DNA column, they are compared to a specimen that could not bind to DNA in the matrix. Stable isotopes are used to identify the specimens. The quantity of the protein complexes is used to identify them (Rep et al., 2002).

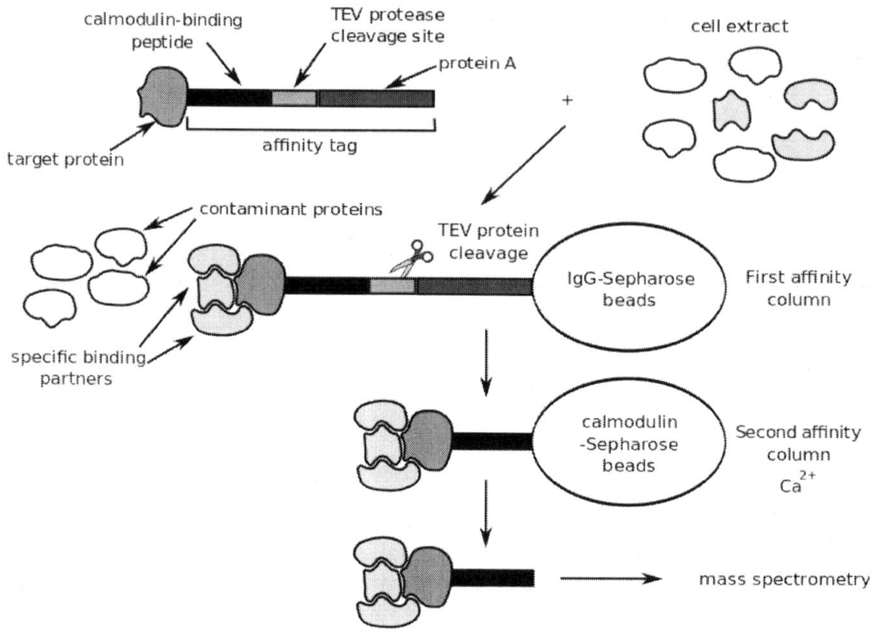

Figure 6.5. Protein-Protein Interactions.

Source: By Philippe Hupé - Emmanuel Barillot, Laurence Calzone, Philippe Hupé, Jean-Philippe Vert, Andrei Zinovyev, Computational Systems Biology of Cancer Chapman & Hall/CRC Mathematical & Computational Biology , 2012, CC BY-SA 3.0, https://commons.wikimedia.org/w/index.php?curid=18532378

6.3.3 Functional Protein Microarray

As said before, a protein array comprises a glass slide about which specific antibodies are put at predetermined places before being subjected to a purified protein. The proteins in the material respond with the antibody at a defined place on the surface, and the antibody is then recognized by interacting with the antibody written in an array at point sets on the slide. Microarray is a technique for determining the relative assortment of various proteins in a type of cell produced under various growth circumstances, or in types of cells from both healthy and diseased people cultivated under the same circumstances (Müller et al., 2001). Protein arrays may also be utilized to look for connections amongst proteins either between enzymes and their substrates or inhibitors. Analytical microarrays, functional microarrays, and reverse-phase microarrays (RPAs) are the three types of protein microarrays now in use. Analytical microarrays are being used to assess the levels of translation and relative abundance of various proteins. Protein binding affinities are also determined using this approach. This is accomplished by exposing a glass slide with antibodies, aptamers, or affibodies printed in fixed places to light (Hu et al., 2011).

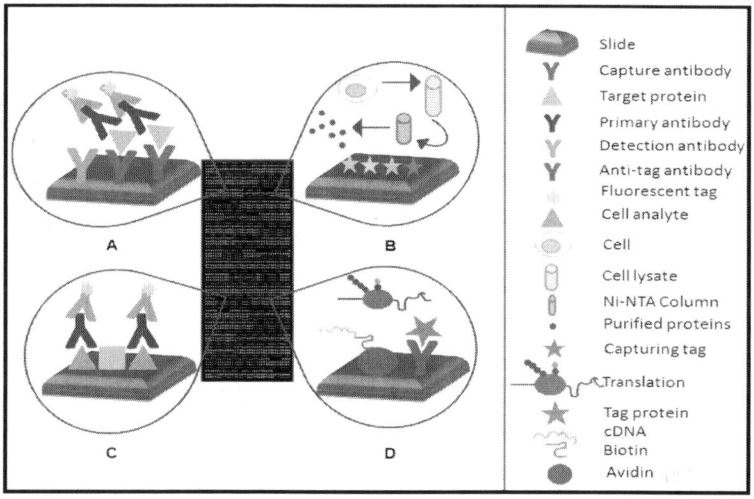

Figure 6.6. Protein-microarrays-a-Functional-protein-microarrays-for-studying-protein: (a) Capture arrays. (b) Cell-based protein microarrays. (c) Reverse phase arrays. (d) Cell-free nucleic acid programmable protein array.

Source: Díez P, Dasilva N, González-González M, Matarraz S, Casado-Vela J, Orfao A, Fuentes M. Data Analysis Strategies for Protein Microarrays. Microarrays. 2012; 1(2):64-83. https://doi.org/10.3390/microarrays1020064

A functioning microarray is a group of fully functioning proteins or protein structures produced on glass slides which are then subjected to protein preparations from a cell which reflects that cell's whole proteome (Bertone & Snyder, 2005). Protein-protein interactions may be determined using this approach. This approach may also be used to anticipate how proteins react with DNA, RNA, phospholipids, and tiny molecules. A cellular protein preparation is fixed on a microscope slide and subsequently probed with a recognized antibody in RPA (Zhou et al., 2012). This approach aids in finding proteins in the proteome of sick cell types that have been changed and cannot bind to a recognized antibody. This approach also determines whether proteins are changed by phosphorylation or other posttranslational alterations in healthy and pathologic situations, as well as under growing conditions. The full yeast protein array, which contains roughly 5800 proteins, was cloned, overexpressed, and put on glass slides to use the functional protein analytical method. GST-His-tags were present on the proteins on the glass plate (Ramachandran et al., 2005). Such labels aided affinity filtration of such proteins on a nickel column before printing on glass slides. By scanning with luminous anti-GST antibody, those labels also assisted in their recognition. In yeast, this research resulted in the discovery of completely undiscovered calmodulin and phospholipid-binding proteins. A similar investigation resulted in the discovery of numerous yeast transmembrane proteins (Ofran&Rost, 2007).

6.4 ANALYSIS OF PROTEIN INTERACTIONS IN SILICO

This method entails doing computer analysis on data from the genome and protein databases. This method compares the relationship of specific nucleic acid sequences, transcribed proteins, or protein structures in various databanks. The most widely used approach is Rosetta Stone, which was created by Marcotte and colleagues (1999). That method is based on the theory that particular network connections form the Rosetta stone protein. The proteins in the Rosetta Stone are found through studying the protein database which is used to build protein networks and interactomes (Marcotte et al., 1999).

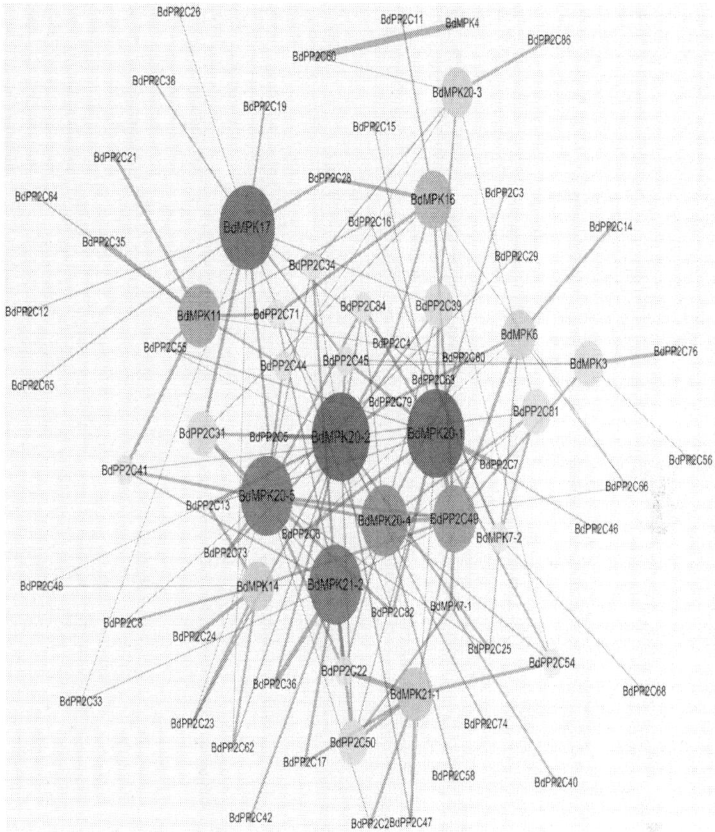

Figure 6.7. Ability To visualize of Protein-Protein Interactions.

Source: Jiang, M., Niu, C., Cao, J. et al. In silico-prediction of protein–protein interactions network about MAPKs and PP2Cs reveals a novel docking site variants in Brachypodium distachyon. Sci Rep 8, 15083 (2018). https://doi.org/10.1038/s41598-018-33428-5

When protein A and protein B show up as single-fused proteins in certain species after scanning the protein database, they might be termed associating proteins. The connecting proteins are shown in the figure below (Murakami et al., 2017).

Structure of Protein A	x-x-x-x-x-x-x-x-x
Structure of Protein B	o-o-o-o-o-o-o-o
Structure of Rosetta Protein/Fused A-B Protein	x-x-x-x-x-x-x-x-x-o-o-o-o-o-o-o-o

The existence of the Rosetta protein in similar animals is searched for using a software program, and if discovered, the Rosetta protein is being used to show that such proteins indeed communicate. The Rosetta protein's results are being used to build the graph or interactome. The web of protein complexes is built using a variety of various computational approaches (Shoemaker &Panchenko, 2007). Among the most significant benefits of computational approaches is that they aid in reducing the high rate of false positives in data generated from yeast two-hybrid investigations. In fact, as explained in detail in this chapter, computational approaches have been employed to analyze the interactomes of various species (Valente et al., 2013). The notion of sequence conservation of protein complexing or interactomes was born out of comparative interactomics. Thus, it has been demonstrated that genes and proteins, and also protein complexes or interactomes in nature, are maintained throughout evolution. Analyzing illnesses, examining the interactomes of easy model organisms like yeast, and undertaking a computer study all benefit from this idea of protein complexes' evolutionary conservation (Valencia &Pazos, 2002). As a result, animal systems may be ignored, and pharmacological adverse effects can be studied in a virtual environment. Statistical approaches have served as the main aim of building PPI maps of various animals due to their speed and capacity to manage a great number of binding proteins at once. Several databanks holding PPI data for various organisms, such as humans, are currently accessible. Statistical approaches have become more popular owing to the accessibility of various protein databanks and software programs (Xing et al., 2016).

6.5 INTERACTOMES

Most proteins occur as complexes; several complexes together can carry out diverse cellular functions in an organism. These functions may involve different metabolic pathways, cell-to-cell communication including signaling, several DNA reactions such as DNA replication, transcription, repair and recombination, cell division and growth. Such a complex of proteins has been called an interactome (Shoemaker &Panchenko, 2007).

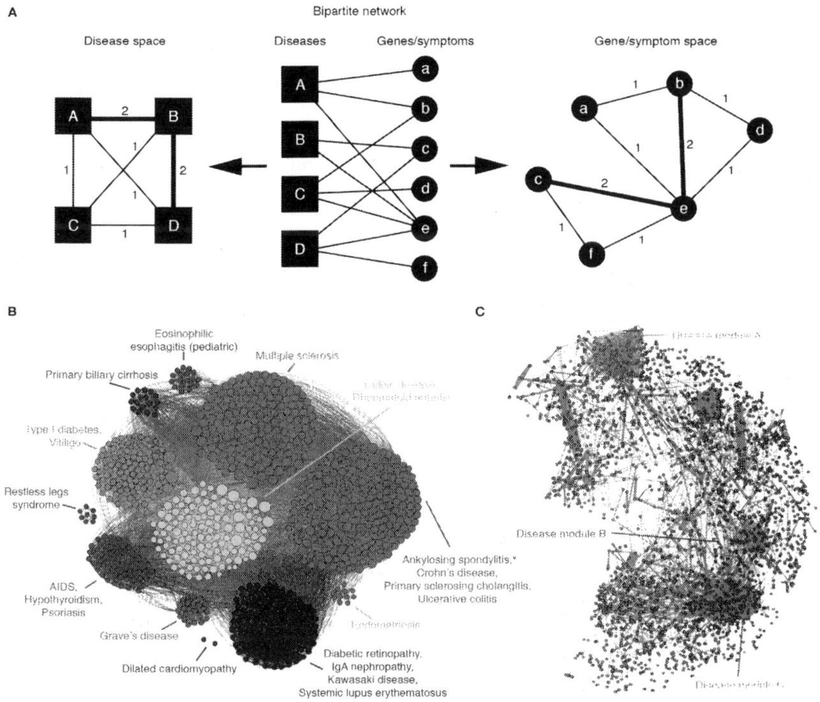

Figure 6.8. 3-D illustration of the Interactome.

Source: Alanis-Lobato G (2015) Mining protein interactomes to improve their reliability and support the advancement of network medicine. Front. Genet. 6:296. doi: 10.3389/fgene.2015.00296

Interactomes or PPI maps have been studied in numerous species, such as viruses, bacteria, and some eukaryotes, such as yeast, worms, flies, mice, humans, and such plants. The majority of such creatures are model organisms whose genetics, chemistry, and molecular genetics are very well (Lambert et al., 2013). The research of interactomes has contributed to the identification of genetic variants and their control. In particular, the analysis of PPI maps offers the foundation for comprehending the diversity of various species, which cannot be described by an easy change in the number of genes. It appears the number of PPIs determines the diversity of a creature. Complex organisms, such as humans, include a greater number of PPIs than yeast, flies, and worms. The research on interactomes has shown that throughout the development of life, not just gene products have been preserved, as well as the interactomes (Isabelle et al., 2010).

Various techniques, including Y2H studies, TPA accompanied by spectrometry, as well as synthetic genetic analysis, have been used to reveal the nature of interactomes. Yeast seems to have around 6000 genes that produce 6000 proteins. Such proteins are temporarily or permanently structured into several interactomes that perform all cellular tasks throughout this organism (Cumberworth et al., 2013). The amount of interactomes varies based on the methodology being used to estimate them and the conclusions of other researchers employing a similar approach. The majority of variance in their quantities is due to false-positive interactions between proteins and the employment of multiple techniques to estimate certain properties. Y2H analysis, for instance, evaluates both temporary and permanent protein-protein interaction, while TAP spectrometry primarily identifies stable protein-protein interaction (Varnaitė &MacNeill, 2016). Moreover, the majority of connections are governed by proteins produced by copying certain genes utilizing polymerase chain reaction (PCR) technology. PCR may produce a shift in the composition of proteins, which might also result in both false-positive and false-negative estimates of protein interactions. Consequently, PCR may result in a substantial difference in the number of interactomes (Das & Yu, 2012).

6.5.1 Prokaryotic Interactomes

PPI maps of numerous bacteria, as well as some viruses, have now been studied in past years. The interactome of *Escherichia coli* is one of those partially understood. *E. coli* is the best-characterized prokaryotic organism in terms of biochemistry, genetics, and bacterial physiology. Starting with the finding of mating in *E. coli*, this individual's early study transformed molecular biology. The research of this species ultimately resulted in the development of molecular cloning techniques using recombinant technologies, ushering in the age of genomics (Bianco, 2021). To use the Y2H test, it has been determined that roughly 4000 prey proteins connect with about 2700 bait proteins in *E. coli*. Seen in these protein-protein interactions are many metabolic processes of the species. *Treponema pallida*, which produces the sexually transmitted illness syphilis in humans, has also been examined for such protein-protein connections. Treponema has one of the shortest bacterial genomes. This organism cannot be cultivated *in vitro* and cannot be analyzed using genetic approaches; consequently, the investigation of protein interactions using the Y2H test has proved useful in comprehending it (Park et al., 2005).

By using Y2H assays, around 1000 proteins have been examined. In Y2H tests, over 700 of such proteins generated over 3600 connections. No syphilis-causing contact has been detected in this research (Klein et al., 2021). Multiple interactomes from different types of bacteria have also been examined. *Helicobacter pylori* is a bacteria that has been widely investigated. This bacteria is present in around fifty percent of human disorders that invade the stomach lining and produce various diseases, including peptic ulcers and sometimes even cancer. Y2H tests have examined more than 260 pylori proteins involved in more than 1200 PPIs. This accounts for 47 percent of its proteome. Nevertheless, a PPI databank is currently developed using statistical approaches (Horkawicz et al., 2014).

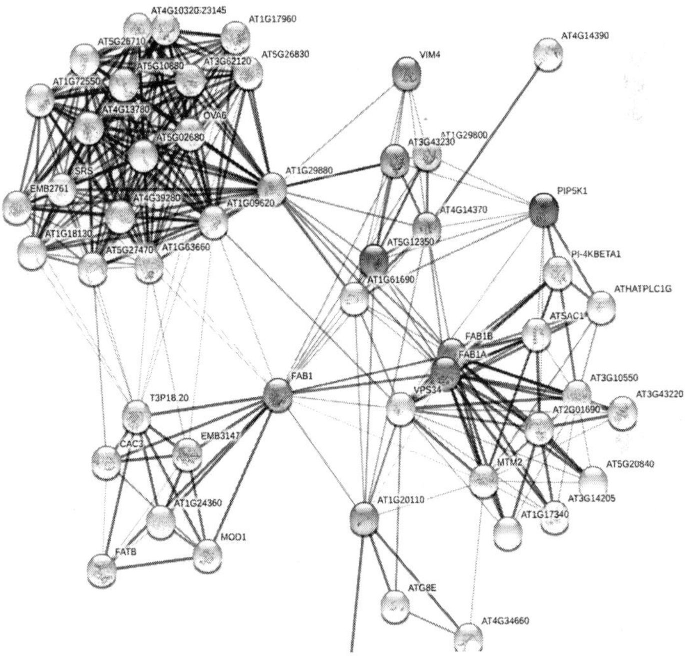

Figure 6.9. Cytoscape depiction of yeast protein-protein contact networks.

Source: Chen Y, Wang W, Liu J, Feng J and Gong X (2020) Protein Interface Complementarity and Gene Duplication Improve Link Prediction of Protein-Protein Interaction Network. Front. Genet. 11:291. doi: 10.3389/fgene.2020.00291

Utilizing Y2H tests, the protein-protein connections of the bacteriophage have been extensively explored in combination with the bacterial interactomes. The majority of these proteins are engaged in the morphogenesis of the bacterium virus (Ding et al., 2020).

6.5.2 Eukaryotic Interactomes

The PPI maps of eukaryotes have been explored. The interactomes of many species, namely flies, yeast, flies, worms, and humans, have been meticulously mapped. Several of those are detailed in the next section (Aw et al., 2016).

6.5.2.1 Yeast Interactome

Yeast is a prototypical easy eukaryote. It has created genetics, biochemistry, and molecular biology departments. It's also the first eukaryote whose whole genomic DNA has been decoded and completely annotated. To gain better information on the system of biochemical processes that govern all cellular functions, such as the genesis, development, and shape of such a species as an instance of the basic eukaryote, the PPI map was tried inside this species (Yu et al., 2008). Various techniques have been employed to construct the web of binding proteins. Such techniques involve spectrometry, bioinformatics, genetic analysis, and Y2H tests. Nearly 40 million protein connections are expected to take place in yeast cells. Only a portion of such connections are shown by some of these methodologies and is the primary reason for the lack of consistency between their results (Tarassov et al., 2008). Other issues plague the outcomes of these methods, including the identification of interactions that may not necessarily exist in the cell. Such data could include false negatives, indicating that the technology is incapable of revealing real cell connections. Statistical examination of the data may allow for the elimination of such.

The PPI map reveals the physical participation of proteins in binary interactions for both the bait protein and the prey protein, as shown by the Y2H test, or in regular connection including all proteins, as indicated by the TAP spectroscopy (Bertin et al., 2007). Two concepts, the spoke system, and the matrix framework are employed to represent the presence of binary and many connections, accordingly. It is claimed that there are three times more binary connections than interactive relationships. The PPI map illustrates that protein interactions take place in a hub or center where multiple proteins connect. Other protein connections link various reaction centers inside this core, suggesting the topology of the interacting proteins system within the cell (Figure 5.3). It has been discovered in several species, such as *E. coli*, yeast, flies, and worms, that proteins with sequence homology in some regions link with one another more often than would be predicted by chance simply (Vo et al., 2016). Pfam group proteins link more often than would be anticipated by chance. The classification of proteins is classified according to the type of regions. A domain is the portion of a protein having a hydrophobic core that has been conserved throughout evolution. Families of proteins are generated depending on the nature of the domain they share; these families are cataloged in various databanks. Pfam is however one database of proteins with many homology domains in their sequence of amino acids. There are now more than 1815 families of Pfam proteins, which account for more than fifty percent of all proteins in eukaryotic organisms. Other groups of protein domains are ProCite, PRINT-S, SMART, and Prodom (Causier, 2004).

6.5.2.2 Fly Interactomes

Drosophila melanogaster is the first multicellular creature for which the whole genome sequence was made accessible. It was the first platform to comprehend multicellularity, tissue/organ formation, and differentiating at the genomic stages. This is an effective paradigm for comprehending human growth and disorders at the most fundamental level of an organization. Utilizing the Y2H test, protein-protein interactions in Drosophila were investigated. An initial map of over 4700 interactions involving over 4600 proteins has been produced. The Drosophila interactome map also reinforces the concept of a system of hub proteins and proteins that link such hubs. There are approximately 65,000 interactions in the Drosophila interactome (Gandhi et al., 2006).

6.5.2.3 Worm Interactome

Caenorhabditis elegans gives a good chance to comprehend the significance of interactomes in determining multicellularity, cell fate, and cell growth throughout the lifetime of a common organism. Combining Y2H tests, TAP spectroscopy, and in silico analysis, *C. elegans* protein interactions are determined using a variety of techniques. Over 3000 proteins directly or indirectly associated with multicellularity have been analyzed. More than 4000 interactions involving diverse biological processes were created by these proteins. Such involves the mechanisms for vulval growth, formation of the germline, pharyngeal functioning, protein breakdown, and DNA repair. The Y2H assays discovered 2900 nodes linked by about 5500 corners within the complex formed (Gandhi et al., 2006). It was discovered that the nodes included ancient, multicellular, and worm-specific groups of proteins. Approximately 700 proteins from an old family were discovered to have yeast orthologs. Around 1100 multicellular proteins were also identified in flies, Arabidopsis, and humans, but over 800 worm-specific proteins were identified (found exclusively in worms). Additionally, it is believed that the worm interactome might well have over 200,000 interactions (Polanska et al., 2009).

As stated before, a knowledge of the PPI map is crucial from several perspectives, including the annotation of gene activity and the involvement of genes and biochemical events in the metabolic regulation and supervision of species. Such an investigation is especially crucial for comprehending human growth and illness. So the protein connection web of model organisms like Escherichia, yeast, flies, and worms was accessible, and attempts were undertaken to create the human PPI map. Several groups' early findings (Seiler & Raul, 2005, Stelzl et al., 2005) showed the existence of a web of protein interactions. And over 3100 new connections comprising 1705 proteins were found by Stelzl et al. These individuals conducted Y2H tests with 5632 preys and 4456 baits. Y2H tests revealed 2,800 interactions in a different Vidal group investigation (Seiler & Raul, 2005). The findings of this investigation were 78 percent identical to those obtained using a different technique, such as TAP spectrometry. Bioinformatic methods have been used to attempt a large-scale determination of the PPI map (Chaurasia et al., 2009); these authors generated an interactive map based on the analysis of data obtained by Y2H assays and other methods, including a statistical approach by different researchers, and designated the network as the unified human interactome (UniH). The UniH has 160,000 unique interactions between 17,000 different human proteins. Currently, UniH is the network map closest to the estimate of 665,000 human interactions (Stumpf et al., 2008) including around 25,000 human genome-encoded proteins.

Plasmodium falciparum is responsible for up to 90 percent of malaria-related fatalities. A concentrated effort has been made to comprehend the genes and proteins of this parasite using genomic and proteomic techniques. Its whole genome has been sequenced, and several genes' roles have been annotated. A pentameric sequence has been found that modifies the parasite and host erythrocyte membrane, hence facilitating parasite entrance. The majority of plasmodium proteins lacked homology with eukaryotic proteins. About eight percent of the total number of genes in the plasmodium have been identified as erythrocyte-targeting proteins, totaling about 400 proteins. The most recent strategy for elucidating the function of different proteins involves the creation of a PPI map, with the expectation that a comprehension of the genes and proteins involved in the pathogenicity of the plasmodium will provide clues and information for the development of antimalarial drugs. Kelley and Idekar demonstrated in 2005 (Suthram et al. 2005 and Sharan et al. 2005) that PPI maps are sufficiently distinct from those of other eukaryotes. Idekar and his team used Y2H tests to investigate over 1300 proteins involved in over 2800 interactions inside the plasmodium. They discovered that the plasmodium has no protein interaction complex with higher eukaryotes such as flies, worms, and humans. However, the plasmodium was shown to share three interaction complexes with yeast (Simonis et al., 2009).

Numerous plant genomes have been completely sequenced, however, their PPI maps are not yet accessible. A beginning has been achieved in the model plant *Arabidopsis thaliana*, which has a smaller genome than other plants such as maize and rice. It has been thoroughly researched to comprehend the molecular biology of plants, especially the genetic control of flowering, nutrient availability, and modification of plants. It is susceptible to manipulation by many genetics as well as molecular biology techniques. This Y2H assays and protoplast two-hybrid system have been utilized to investigate the binary interplay between many proteins in this organism. That using these approaches, plant gene encoding has been thoroughly explored. Using statistical approaches, a dataset of the PPI in Arabidopsis has been produced (Mallam &Marcotte, 2017).

6.5.2.4 Interactome During Human Development and Disease

Human evolution starts with the creation of an embryo, which would be the consequence of cell growth and division, resulting in the production of over 700 types of cells more than a billion cells as just an adult. Due to the obvious varied genetic expression that describes the function and structure of a given cell, every cell type has a diverse proteome; hence, a neuron cell is unique from a muscle cell. The neuronal cell and the muscle fiber have wholly distinct protein profiles which provide them their distinct structure and function. Several proteins distinguish these two cell types. Just several but distinct proteins are produced by a variety of cell types. The beta cells (pancreas), for instance, mostly produce insulin, while erythrocytes virtually solely produce hemoglobin. The biochemical/molecular foundation of human growth and function is being attempted to describe the links of distinct types of cells. Although not much work has been done in that regard, examining the protein composition of human blastocytes at various phases of development provides a start. In the lab of Mandy Katz-Jaffe, blastocysts produced from discarded embryos received from an *in vitro* fertilization clinic in Colorado were evaluated for their protein profiles using gel electrophoresis (2006).

The concentration of numerous proteins differs in early and advanced blastocysts, according to this study. Early embryogenesis is attributed to several biomarkers, including parathyroid hormone-related peptide growth factor-like epidermal growth factor. Since of ethical issues and regulations, there are a limited number of human embryos available for research. Once it becomes feasible to examine stem cells, it is thought that certain progress may be achieved in this area. As a result, mapping the interactome of a growing human embryo is now impossible. However, research in model animals like mice and pigs is thought to give PPI network information that is relevant to human evolution (Katz-Jaffe et al., 2007).

6.6 EVOLUTION AND CONSERVATION OF INTERACTIONS

All in our cosmos seems to have evolved, and this is especially true of biological systems. It seems that evolution operates by adapting current infrastructure to meet the demands of organisms in specific environments. This is true of the PPI map as well. The concept of interactome conservation is founded on the notion of interacting proteins coevolving. The interactomes have been preserved in living systems, according to Marc Vidal. Ideker, on the other hand, was the one who produced the proof. Idekar et al. (2001) studied protein interaction networks spanning 6000 proteins in yeast, flies, worms, and humans.

These researchers discovered that 71 connections were maintained throughout 3 varieties: flies, worms, and humans. Their findings were based on the examination of many interactions between proteins with known and unknown activities. The attribution of function(s) to proteins that could not be anticipated only based on homology was a key discovery of their research. This team also discovered that Plasmodium falciparum (the parasite that produces malaria in humans) has no interaction route in common with just about any higher eukaryote.

Geneticists have been confronted with a problem known as the "C-value paradox" from the start of the chromosomal theory of inheritance. That is, the number of genes in an organism did not correspond with its complexity. Humans, for example, have roughly 24,000 genes, while worms have about 19,000 genes and fruit flies have 14,000 genes. Stumpf (2008) stated that the number of protein interactions dictates the complexity of an organism based on a recent investigation of the number of protein interactions in various animals. Humans, for example, have ten times as many protein interactions as fruit flies and three times as many as worms. In their research, Stumpf et al. (2008) only looked at humans, fruit flies, and worms. When considering the number of protein interactions in yeast and fruit flies, it appears that neither the quantity as well as the kind of contacts must be addressed when understanding an organism's diversity. The research of plants like maize and rice may provide a lot of evidence for this theory about the relevance of the series of exchanges in understanding the complexity of organisms such as humans. Some plants have genomes that are bigger than humans' yet significantly less complicated. As a result, the number of protein interactions in maize and rice should be fewer than that expected for humans. These statistics are temporarily unavailable.

6.7 INTERACTION OF PROTEINS WITH SMALL MOLECULES

In each species, protein-protein interactions regulate cellular structure and function. Molecules often connect with proteins to aid their functioning or, on rare occasions, to disturb these connections. Cofactors, coenzymes, inhibitors, stabilizers, and allosteric effectors are the most common chemicals that help proteins operate. More than 4000 tiny compounds interact with more than 20,000 proteins, according to a study of the protein databank. Metal ions including magnesium, calcium, and zinc, as well as sugar molecules and nucleotides, make up the majority of these. In general, small compounds bind with a specific amino acid in a protein (Vidal et al., 2011).

REFERENCES

1. Aw, J. G. A., Shen, Y., Wilm, A., Sun, M., Lim, X. N., Boon, K. L., ... & Wan, Y. (2016). *In vivo* mapping of eukaryotic RNA interactomes reveals principles of higher-order organization and regulation. *Molecular cell*, *62*(4), 603-617.
2. Azzazy, H. M., & Highsmith Jr, W. E. (2002). Phage display technology: clinical applications and recent innovations. *Clinical biochemistry*, *35*(6), 425-445.
3. Bartel, P., Chien, C. T., Sternglanz, R., & Fields, S. (1993). Elimination of false positives that arise in using the two-hybrid system. *Biotechniques*, *14*(6), 920-924.
4. Bertin, N., Simonis, N., Dupuy, D., Cusick, M. E., Han, J. D. J., Fraser, H. B., ... & Vidal, M. (2007). Confirmation of organized modularity in the yeast interactome. *PLoS biology*, *5*(6), 153.
5. Bertone, P., & Snyder, M. (2005). Advances in functional protein microarray technology. *The FEBS journal*, *272*(21), 5400-5411.
6. Bianco, P. R. (2021). The mechanism of action of the SSB interactome reveals it is the first OB-fold family of genome guardians in prokaryotes. *Protein Science*, *30*(9), 1757-1775.
7. Boy-Marcotte, E., Lagniel, G., Perrot, M., Bussereau, F., Boudsocq, A., Jacquet, M., &Labarre, J. (1999). The heat shock response in yeast: differential regulations and contributions of the Msn2p/Msn4p and Hsf1p regulons. *Molecular microbiology*, *33*(2), 274-283.
8. Cafarelli, T. M., Desbuleux, A., Wang, Y., Choi, S. G., De Ridder, D., & Vidal, M. (2017). Mapping, modeling, and characterization of protein–protein interactions on a proteomic scale. *Current Opinion in Structural Biology*, *44*, 201-210.
9. Causier, B. (2004). Studying the interactome with the yeast two-hybrid system and mass spectrometry. *Mass spectrometry reviews*, *23*(5), 350-367.
10. Causier, B., & Davies, B. (2002). Analyzing protein-protein interactions with the yeast two-hybrid system. *Plant molecular biology*, *50*(6), 855-870.
11. Chaurasia, S. S., Kaur, H., de Medeiros, F. W., Smith, S. D., & Wilson, S. E. (2009). Dynamics of the expression of intermediate filaments vimentin and desmin during myofibroblast differentiation after corneal injury. *Experimental eye research*, *89*(2), 133-139.

12. Chua, H. N., & Wong, L. (2008). Increasing the reliability of protein interactomes. *Drug discovery today*, *13*(15-16), 652-658.
13. Cumberworth, A., Lamour, G., Babu, M. M., &Gsponer, J. (2013). Promiscuity as a functional trait: intrinsically disordered regions as central players of interactomes. *Biochemical Journal*, *454*(3), 361-369.
14. Das, J., & Yu, H. (2012). HINT: High-quality protein interactomes and their applications in understanding human disease. *BMC systems biology*, *6*(1), 1-12.
15. Ding, W., Tan, H. Y., Zhang, J. X., Wilczek, L. A., Hsieh, K. R., Mulkin, J. A., & Bianco, P. R. (2020). The mechanism of Single strand binding protein–RecG binding: Implications for SSB interactome function. *Protein Science*, *29*(5), 1211-1227.
16. Doi, N., Takashima, H., Kinjo, M., Sakata, K., Kawahashi, Y., Oishi, Y., ... &Yanagawa, H. (2002). Novel fluorescence labeling and high-throughput assay technologies for *in vitro* analysis of protein interactions. *Genome Research*, *12*(3), 487-492.
17. Dunin-Horkawicz, S., Kopec, K. O., &Lupas, A. N. (2014). Prokaryotic ancestry of eukaryotic protein networks mediating innate immunity and apoptosis. *Journal of molecular biology*, *426*(7), 1568-1582.
18. Fields, S. (2005). High-throughput two-hybrid analysis: The promise and the peril. *The FEBS journal*, *272*(21), 5391-5399.
19. Fields, S., & Song, O. K. (1989). A novel genetic system to detect protein–protein interactions. *Nature*, *340*(6230), 245-246.
20. Galarneau, A., Primeau, M., Trudeau, L. E., &Michnick, S. W. (2002). β-Lactamase protein fragment complementation assays as *in vivo* and *in vitro* sensors of protein–protein interactions. *Nature Biotechnology*, *20*(6), 619-622.
21. Gandhi, T. K. B., Zhong, J., Mathivanan, S., Karthick, L., Chandrika, K. N., Mohan, S. S., ... & Pandey, A. (2006). Analysis of the human protein interactome and comparison with yeast, worm and fly interaction datasets. *Nature genetics*, *38*(3), 285-293.
22. Goeury, K., Duy, S. V., Munoz, G., Prévost, M., & Sauvé, S. (2019). Analysis of Environmental Protection Agency priority endocrine disruptor hormones and bisphenol A in tap, surface and wastewater by online concentration liquid chromatography-tandem mass spectrometry. *Journal of Chromatography A*, *1591*, 87-98.

23. Hu, S., Xie, Z., Qian, J., Blackshaw, S., & Zhu, H. (2011). Functional protein microarray technology. *Wiley Interdisciplinary Reviews: Systems Biology and Medicine*, *3*(3), 255-268.
24. Ideker, T., Galitski, T., & Hood, L. (2001). A new approach to decoding life: systems biology. *Annual review of genomics and human genetics*, *2*(1), 343-372.
25. Isabelle, M., Moreel, X., Gagné, J. P., Rouleau, M., Ethier, C., Gagné, P., ... & Poirier, G. G. (2010). Investigation of PARP-1, PARP-2, and PARG interactomes by affinity-purification mass spectrometry. *Proteome science*, *8*(1), 1-11.
26. James, P., Halladay, J., & Craig, E. A. (1996). Genomic libraries and a host strain designed for highly efficient two-hybrid selection in yeast. *Genetics*, *144*(4), 1425-1436.
27. Jauregui, O., Moyano, E., & Galceran, M. T. (1997). Liquid chromatography-atmospheric pressure ionization mass spectrometry for the determination of chloro-and nitrophenolic compounds in tap water and sea water. *Journal of Chromatography A*, *787*(1-2), 79-89.
28. Katz-Jaffe, M. G., Schoolcraft, W. B., & Gardner, D. K. (2006). Analysis of protein expression (secretome) by human and mouse preimplantation embryos. *Fertility and sterility*, *86*(3), 678-685.
29. Kawahashi, Y., Doi, N., Takashima, H., Tsuda, C., Oishi, Y., Oyama, R., ... &Yanagawa, H. (2003). *In vitro* protein microarrays for detecting protein-protein interactions: Application of a new method for fluorescence labeling of proteins. *PROTEOMICS: International Edition*, *3*(7), 1236-1243.
30. Kelley, R., &Ideker, T. (2005). Systematic interpretation of genetic interactions using protein networks. *Nature Biotechnology*, *23*(5), 561-566.
31. Klein, B., Hoel, E., Swain, A., Griebenow, R., & Levin, M. (2021). Evolution and emergence: higher order information structure in protein interactomes across the tree of life. *Integrative Biology*, *13*(12), 283-294.
32. Lambert, J. P., Ivosev, G., Couzens, A. L., Larsen, B., Taipale, M., Lin, Z. Y., ... &Gingras, A. C. (2013). Mapping differential interactomes by affinity purification coupled with data-independent mass spectrometry acquisition. *Nature methods*, *10*(12), 1239-1245.

33. Lentze, N., &Auerbach, D. (2008). Membrane-based yeast two-hybrid system to detect protein interactions. *Current Protocols in Protein Science*, *52*(1), 19-17.

34. Luban, J., & Goff, S. P. (1995). The yeast two-hybrid system for studying protein—protein interactions. *Current opinion in biotechnology*, *6*(1), 59-64.

35. Luo, Y., Batalao, A., Zhou, H., & Zhu, L. (1997). Mammalian two-hybrid system: a complementary approach to the yeast two-hybrid system. *Biotechniques*, *22*(2), 350-352.

36. Mallam, A. L., &Marcotte, E. M. (2017). Systems-wide studies uncover commander, a multiprotein complex essential to human development. *Cell systems*, *4*(5), 483-494.

37. Montiel-León, J. M., Duy, S. V., Munoz, G., Amyot, M., & Sauvé, S. (2018). Evaluation of on-line concentration coupled to liquid chromatography tandem mass spectrometry for the quantification of neonicotinoids and fipronil in surface water and tap water. *Analytical and bioanalytical chemistry*, *410*(11), 2765-2779.

38. Müller, D. R., Schindler, P., Towbin, H., Wirth, U., Voshol, H., Hoving, S., & Steinmetz, M. O. (2001). Isotope-tagged cross-linking reagents. A new tool in mass spectrometric protein interaction analysis. *Analytical chemistry*, *73*(9), 1927-1934.

39. Murakami, Y., Tripathi, L. P., Prathipati, P., &Mizuguchi, K. (2017). Network analysis and in silico prediction of protein–protein interactions with applications in drug discovery. *Current opinion in structural biology*, *44*, 134-142.

40. Nero, T. L., Morton, C. J., Holien, J. K., Wielens, J., & Parker, M. W. (2014). Oncogenic protein interfaces: small molecules, big challenges. *Nature Reviews Cancer*, *14*(4), 248-262.

41. Ofran, Y., &Rost, B. (2007). Protein–protein interaction hotspots carved into sequences. *PLoS computational biology*, *3*(7), 119.

42. Park, D., Lee, S., Bolser, D., Schroeder, M., Lappe, M., Oh, D., &Bhak, J. (2005). Comparative interactomics analysis of protein family interaction networks using PSIMAP (protein structural interactome map). *Bioinformatics*, *21*(15), 3234-3240.

43. Piehler, J. (2005). New methodologies for measuring protein interactions *in vivo* and *in vitro*. *Current opinion in structural biology*, *15*(1), 4-14.

44. Polanska, U. M., Fernig, D. G., &Kinnunen, T. (2009). Extracellular interactome of the FGF receptor–ligand system: Complexities and the relative simplicity of the worm. *Developmental Dynamics*, *238*(2), 277-293.
45. Ramachandran, N., Larson, D. N., Stark, P. R., Hainsworth, E., &LaBaer, J. (2005). Emerging tools for real-time label-free detection of interactions on functional protein microarrays. *The FEBS journal*, *272*(21), 5412-5425.
46. Raman, K. (2010). Construction and analysis of protein–protein interaction networks. *Automated experimentation*, *2*(1), 1-11.
47. Rep, M., Dekker, H. L., Vossen, J. H., de Boer, A. D., Houterman, P. M., Speijer, D., ... & Cornelissen, B. J. (2002). Mass spectrometric identification of isoforms of PR proteins in xylem sap of fungus-infected tomato. *Plant Physiology*, *130*(2), 904-917.
48. Rual, J. F., Venkatesan, K., Hao, T., Hirozane-Kishikawa, T., Dricot, A., Li, N., ... & Vidal, M. (2005). Towards a proteome-scale map of the human protein–protein interaction network. *Nature*, *437*(7062), 1173-1178.
49. Sato, T., Hanada, M., Bodrug, S., Irie, S., Iwama, N., Boise, L. H., ... & Wang, H. G. (1994). Interactions among members of the Bcl-2 protein family analyzed with a yeast two-hybrid system. *Proceedings of the National Academy of Sciences*, *91*(20), 9238-9242.
50. Seiler, N., & Raul, F. (2005). Polyamines and apoptosis. *Journal of cellular and molecular medicine*, *9*(3), 623-642.
51. Serebriiskii, I., Estojak, J., Berman, M., &Golemis, E. A. (2000). Approaches to detecting false positives in yeast two-hybrid systems. *Biotechniques*, *28*(2), 328-336.
52. Sharan, R., Ideker, T., Kelley, B., Shamir, R., & Karp, R. M. (2005). Identification of protein complexes by comparative analysis of yeast and bacterial protein interaction data. *Journal of computational biology*, *12*(6), 835-846.
53. Shoemaker, B. A., &Panchenko, A. R. (2007). Deciphering protein–protein interactions. Part II. Computational methods to predict protein and domain interaction partners. *PLoS computational biology*, *3*(4), e43.
54. Simonis, N., Rual, J. F., Carvunis, A. R., Tasan, M., Lemmens, I., Hirozane-Kishikawa, T., ... & Vidal, M. (2009). Empirically controlled

mapping of the Caenorhabditiselegans protein-protein interactome network. *Nature methods*, *6*(1), 47-54.

55. Smith, G. P., &Petrenko, V. A. (1997). Phage display. *Chemical reviews*, *97*(2), 391-410.
56. Stelzl, U., Worm, U., Lalowski, M., Haenig, C., Brembeck, F. H., Goehler, H., ... & Wanker, E. E. (2005). A human protein-protein interaction network: a resource for annotating the proteome. *Cell*, *122*(6), 957-968.
57. Stumpf, M. P., Thorne, T., De Silva, E., Stewart, R., An, H. J., Lappe, M., &Wiuf, C. (2008). Estimating the size of the human interactome. *Proceedings of the National Academy of Sciences*, *105*(19), 6959-6964.
58. Suthram, S., Sittler, T., &Ideker, T. (2005). The Plasmodium protein network diverges from those of other eukaryotes. *Nature*, *438*(7064), 108-112.
59. Takaku, Y., Shimamura, T., Masuda, K., & Igarashi, Y. (1995). Iodine determination in natural and tap water using inductively coupled plasma mass spectrometry. *Analytical sciences*, *11*(5), 823-827.
60. Tarassov, K., Messier, V., Landry, C. R., Radinovic, S., Molina, M. M. S., Shames, I., ... &Michnick, S. W. (2008). An *in vivo* map of the yeast protein interactome. *Science*, *320*(5882), 1465-1470.
61. Valencia, A., &Pazos, F. (2002). Computational methods for the prediction of protein interactions. *Current opinion in structural biology*, *12*(3), 368-373.
62. Valente, G. T., Acencio, M. L., Martins, C., & Lemke, N. (2013). The development of a universal in silico predictor of protein-protein interactions. *PloS one*, *8*(5), 65587.
63. Varnaitė, R., &MacNeill, S. A. (2016). Meet the neighbors: Mapping local protein interactomes by proximity-dependent labeling with BioID. *Proteomics*, *16*(19), 2503-2518.
64. Vidal, M., Cusick, M. E., &Barabási, A. L. (2011). Interactome networks and human disease. *Cell*, *144*(6), 986-998.
65. Vo, T. V., Das, J., Meyer, M. J., Cordero, N. A., Akturk, N., Wei, X., ... & Yu, H. (2016). A proteome-wide fission yeast interactome reveals network evolution principles from yeasts to human. *Cell*, *164*(1-2), 310-323.

66. Walhout, A. J., & Vidal, M. (2001). High-throughput yeast two-hybrid assays for large-scale protein interaction mapping. *Methods*, *24*(3), 297-306.

67. Weinzierl, A. O., Rudolf, D., Hillen, N., Tenzer, S., van Endert, P., Schild, H., ... &Stevanović, S. (2008). Features of TAP-independent MHC class I ligands revealed by quantitative mass spectrometry. *European journal of immunology*, *38*(6), 1503-1510.

68. Xing, S., Wallmeroth, N., Berendzen, K. W., &Grefen, C. (2016). Techniques for the analysis of protein-protein interactions in vivo. *Plant physiology*, *171*(2), 727-758.

69. Yu, H., Braun, P., Yıldırım, M. A., Lemmens, I., Venkatesan, K., Sahalie, J., ... & Vidal, M. (2008). High-quality binary protein interaction map of the yeast interactome network. *Science*, *322*(5898), 104-110.

70. Zhou, S. M., Cheng, L., Guo, S. J., Zhu, H., & Tao, S. C. (2012). Functional protein microarray: an ideal platform for investigating protein binding property. *Frontiers in biology*, *7*(4), 336-349.

CHAPTER 7

APPLICATIONS OF PROTEOMICS

CONTENTS

7.1 INTRODUCTION

As mentioned throughout this chapter, proteomics technologies have been widely employed to comprehend human disorders and are the foundation of medicine in easing such diseases (Petricoin&Liotta, 2003).

Gorrod (1909), in his classic 1903 paper on inborn defects of metabolism, was the first to claim that genes govern illnesses. He documented the inheritance of alcaptonuria and numerous other human disorders in this work. The one-gene–one-enzyme hypothesis proposed by Beadle and Tatum in 1941 postulated a relationship between genes, proteins, and illnesses, and also the possibility of medical assistance in disease. This idea proposed that in a person with a heritable condition, a specific protein governing cell shape or a chemical response is either absent or faulty (Mouradian, 2002).

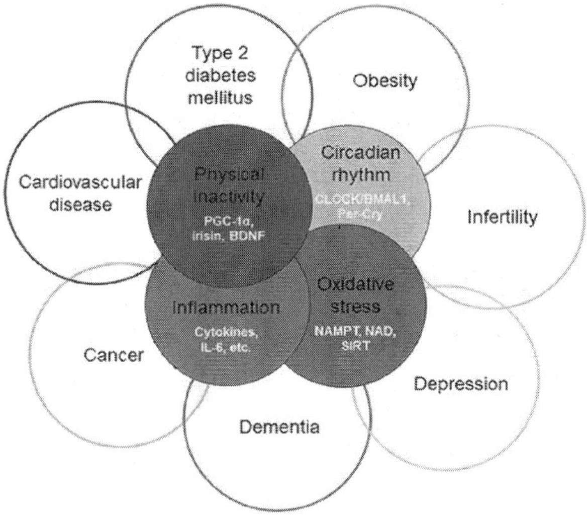

Figure 7.1. Applications of Proteomics.

Source: Zsuga J, More CE, Erdei T, Papp C, Harsanyi S and Gesztelyi R (2018) Blind Spot for Sedentarism: Redefining the Diseasome of Physical Inactivity in View of Circadian System and the Irisin/BDNF Axis. Front. Neurol. 9:818. doi: 10.3389/fneur.2018.00818

The finding that such hemoglobin protein is changed in sickle cell anemia patients provided the initial evidence for this viewpoint. Following that, numerous human disorders were discovered to have a faulty protein or to be lacking that protein entirely. As a result, the one-gene–one-enzyme paradigm formed the foundation of medical therapy by delivering insulin, which is lacking in diabetes patients, or by providing the end product of a biochemical process that is not being created by the deficient enzymatic activity in an afflicted person (for instance, a thyroxin medicine was given to patients suffering from thyroid diseases) (Yokota, 2019).

7.2 DISEASOME

In 2007, Marc Vidal and his team at Harvard University envisioned the diseasome as a system of human disorders. It seeks to create a relationship between gene (disease genome) and abnormality (disease phenome) in people. Numerous human illnesses are monogenic, meaning that they are regulated by one gene. In certain cases, though, mutations at various places in a similar gene result in diseases with distinct symptoms (Goh & Choi, 2012). TP53 mutations, for instance, have been linked to 11 distinct cancer phenotypes. Zellweger syndrome, on the other hand, may be caused by mutations in any of eleven distinct genes. Vidal 2007 (see Goh et al. 2007) established the notion of diseasome in the shape of a gene-disease graph to illustrate all of these instances of the relationship between the disease genome and illness phenome (Midic et al., 2009). In this figure, the one-gene– the one-illness relationship was associated with a single line linking the gene to the disease, but the other gene-disease relationships were seen by separate linking a disease to many genes or, inversely, one gene to multiple diseases. That prototype of the diseasome is accompanied by a set of factors, such as the genetic regulation of illness, the fundamental protein-protein connections outlined in an earlier chapter, and the interconnected nature of cellular metabolism. Figure 6.1 depicts the structure of the diseasome (Wysocki& Ritter, 2011).

Vidal and many others have investigated 1284 gene-disease associations. In minimum 867 instances, they have demonstrated that one gene is associated with one illness. In 516 instances, though, an illness was regulated by much more than a gene, such that numerous diseases had a few similar components. Goh et al. (2007) discovered that complicated illnesses, such as cancer, cardiovascular disease, and neurological disorders, were diverse, regulated by multiple genes, and linked to certain genes, while metabolic disease is often ruled by a centralized gene (Barabási, 2007).

7.3 MEDICAL PROTEOMICS

Medical proteomics tries to characterize specific proteins since proteins are useful in medicine as therapeutics or as targets for drug research, and the volume of the rings and rectangles on their edges is equivalent to the number of genes and diseases associated (Cash, 2000). Therefore, under the guidance and funding of HUPO, extensive efforts are underway to explain the function and structure of proteins, and also their interconnections, to assess their involvement in the formation of novel medications, and to comprehend and mitigate their harmful side effects. Proteomics of many human components is now being investigated. Under the funding of HUPO, the serum proteome is being explored in the United States, the liver proteome has been researched in China, and the brain proteome has been researched in Germany. Various labs are looking into various proteomes, such as the cancer proteome and the cardiovascular proteome (Cuervo et al., 2010).

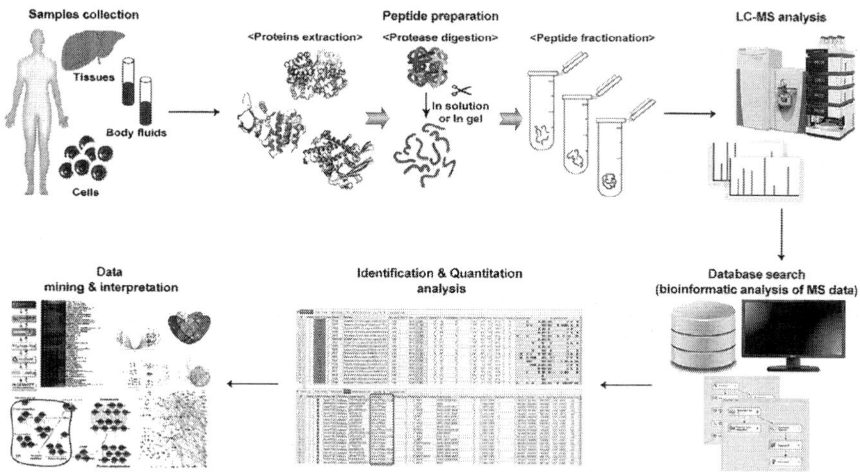

Figure 7.2. Proteomics' Part in the Growth of Personalized Cancer Medicine.

Source: Kwon YW, Jo H-S, Bae S, Seo Y, Song P, Song M and Yoon JH (2021) Application of Proteomics in Cancer: Recent Trends and Approaches for Biomarkers Discovery. Front. Med. 8:747333. doi: 10.3389/fmed.2021.747333

7.3.1 Body Fluid Proteome

Proteomics of bodily fluids is significant since body fluids move thru the body and get into touch with multiple tissues of various organs; therefore, it is feasible to detect proteins that could serve as perfect biomarkers of illness and give hints for drug discovery (Huang et al., 2021). Blood (including plasma and serum), urine, cerebrospinal fluid, amniotic fluid, salivary gland secretion, nasal secretion, tears, and nasopharyngeal aspirate have been investigated in depth. The significant differences in the quantity of a single protein between persons and between specimens obtained at various periods from a similar individual are the greatest technological barrier in the field of proteomics of bodily fluids. To define a baseline level for a given protein, it is necessary to examine a large number of people at varied periods or settings (Hu et al., 2006).

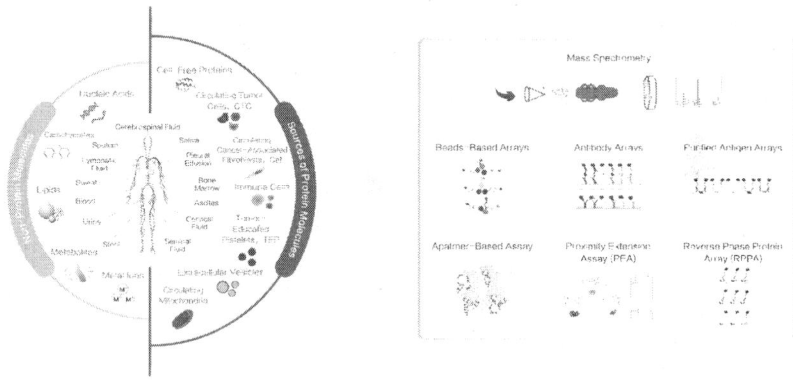

Figure 7.3. Fluids in the Body.

Source: Ding, Z., Wang, N., Ji, N. et al. Proteomics technologies for cancer liquid biopsies. Mol Cancer 21, 53 (2022). https://doi.org/10.1186/s12943-022-01526-8

7.3.2 Liver Proteome

From a biological, physiological, pathological, and pharmacological standpoint, the liver is a vital organ. In regards to length and diversity, it is only next to the mind. It regulates the digestive process, embryonic red blood cell production, immunological feature, and xenobiotic detoxification in the organism. Furthermore, it creates retinol. It contains additional proteins which attach to medications and help them move about the body, which would be useful in pharmacological research (Shi et al., 2007). The liver's metabolic activities and biliary secretion also aid in drug detoxification. From a medical standpoint, the liver is vulnerable to liver cancer, alcohol-induced liver cirrhosis, and viral hepatitis, which particularly affects globally. The liver proteome has been studied in-depth in several labs, mostly in China under the support of HUPO, due to its essential role in biology, pathophysiology, and drug metabolism (Ding et al., 2016).

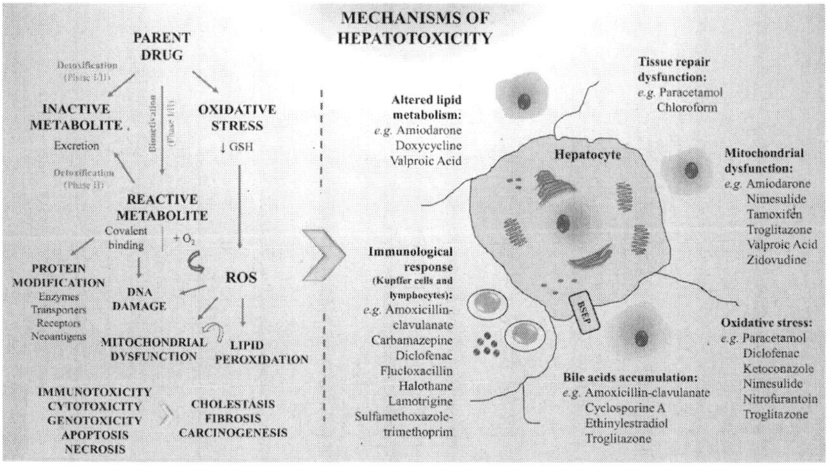

Figure 7.4. Hepatocytes' secretory protein processing.

Source: Serras AS, Rodrigues JS, Cipriano M, Rodrigues AV, Oliveira NG and Miranda JP (2021) A Critical Perspective on 3D Liver Models for Drug Metabolism and Toxicology Studies. Front. Cell Dev. Biol. 9:626805. doi: 10.3389/fcell.2021.626805

This research looked at the proteomes of mature individuals, fetal, and diseased livers from a variety of individuals. Or more than 5000 distinct proteins have been found as part of the HUPO liver proteome research. Several proteins are all being investigated as potential indicators of liver disease. A thorough examination of the fetal liver revealed 2495 unique proteins (He, 2005). Proteomic studies of fetal, adult, and cancer liver cells were also performed. Proteins isolated from various liver cells produced in cultured cells were used in this research. Human fetal hepatocytes (HFH), human hepatocytes (HH4), and human liver cancer cells (Huh7) were among the cell lines used. A total of 2159 distinct proteins were discovered in such cells, with 496 of them being detected throughout all three cell lines. Yan et al. discovered features of diverse cell lines like Huh7, HFH, and HH4 cells between these 337, 364, and 414 proteins, respectively (2004). Furthermore, the mouse liver has been created as a model system for proteomic research. Several 3244 distinct proteins have been discovered in mouse liver, with 47 percent being membrane-bound and roughly 35 percent containing transmembrane peptides (Kuscuoglu et al., 2018).

7.3.3 Brain Proteome

HUPO has allocated the Brain Proteome Project principally to Germany, but it is also being studied in labs in other nations. The primary goal of the brain proteome study was to understand all the proteins in the mind and to discover the proteins implicated in various neurodegenerative illnesses, like Alzheimer's and Parkinson's. Nonetheless, the Human Brain Project has achieved little development so far, except the normalization of many protein stages, which will assist detect the variations in protein levels across patients and create specific biomarkers for developing drugs and probable illness therapy (Klose et al., 2002).

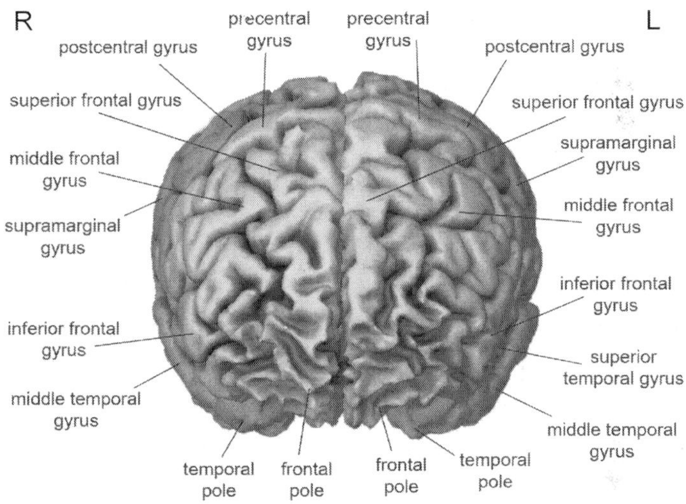

Figure 7.5. Human Brain Structure.

Source: Georgiev DD, Georgieva I, Gong Z, Nanjappan V, Georgiev GV. Virtual Reality for Neurorehabilitation and Cognitive Enhancement. Brain Sciences. 2021; 11(2):221. https://doi.org/10.3390/brainsci11020221

Other than that of the Human Brain Project, the mouse brain proteome has been designated as a physical model. The mouse brain proteome has indeed been extensively described. There are at minimum 7792 proteins in the mouse proteome (Wang et al., 2007), and 1564 of them are recognized as cysteinyl peptides. Approximately 26% of proteins were identified as membrane proteins having a role in transport and cell signaling. Over 1400 proteins were discovered to include transmembrane domains (Sharma et al., 2015).

7.3.4 Heart/Cardiovascular Proteome

Proteins define a cell's function and structure, and consequently an organism's. Throughout the differentiating and developmental stages of an organ's existence, proteins are prone to alter based on cellular activity. Proteins may also alter under pathological situations, either as a source or as a result of the illness. Proteins may also vary in reaction to drugs and other regular activities, as well as any other environmental variables including temperature, nutrition, and allergies. The proteome of the heart and circulatory system has therefore been studied to detect heart disease signs and develop medications to cure them (HW Dekkers et al., 2010). A protein map of the human heart's left ventricle has been created using a 2D gel study of the cardiac proteome. The proteome of the left ventricle contains about 110 distinct proteins, according to this protein atlas. In rats, the influence of exercise on the cardiac proteome was studied (Sung et al., 2006). When compared to inactive rats, animals that undertook an exercise program revealed changes in 26 proteins. Many proteins and changes in such proteins have been examined as results of the human heart and vascular network using spectrometric analysis following 2D gel or other protein isolation. The human myocardium has been shown to have around 200 proteins. Proteins from the sarcoplasmic/endoplasmic reticulum have also been discovered (Channaveerappa et al., 2019). Heat shock proteins (Hsps) and mitochondrial proteins associated with energy generation were found to have lower amounts in heart disease patients (Arrell et al. 2008). Several proteins like crystalline, HSP27, and actin have their levels raised. Under illness circumstances, the heart demonstrated variations in the transcription of distinct isoforms of the enzyme/protein or posttranslational alteration of proteins, adding to variations in the size of particular proteins (Pedret et al., 2021).

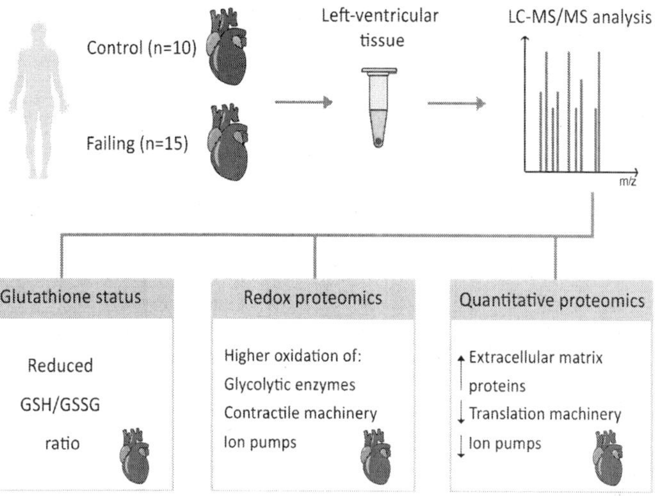

Figure 7.6. Proteomics of the Human Heart.

Source: Tomin T, Schittmayer M, Sedej S, Bugger H, Gollmer J, Honeder S, Darnhofer B, Liesinger L, Zuckermann A, Rainer PP, Birner-Gruenberger R. Mass Spectrometry-Based Redox and Protein Profiling of Failing Human Hearts. International Journal of Molecular Sciences. 2021; 22(4):1787. https://doi.org/10.3390/ijms22041787

7.4 CLINICAL PROTEOMICS

As mentioned below, proteomics will play a significant role in the prevention and therapy of human illnesses in the history of medicine (Liotta et al., 2003).

7.4.1 Genomics and Proteomics of Human Diseases

Genetics and genomics have aided in establishing the reality of human illnesses and their hereditary or nongenetic causes. To establish the genetic basis of an illness, traditional genetics and epidemiology or population genetic approaches have been applied. Disease-causing genes, their chromosomal placements, and DNA pattern, as well as the proteins regulated by such genes, have been found utilizing genetics and genomics. The chromosomal position was determined utilizing somatic cell genetics, cytogenetics, and translocation roadmap (Macaulay et al., 2005).

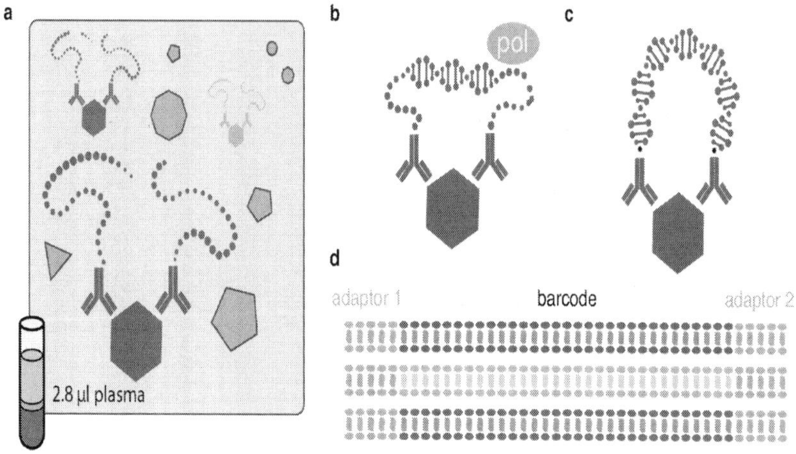

Figure 7.7. The PEA-NGS method's principle and cohort descriptions.

Source: Zhong, W., Edfors, F., Gummesson, A. et al. Next generation plasma proteome profiling to monitor health and disease. Nat Commun 12, 2493 (2021). https://doi.org/10.1038/s41467-021-22767-z

Genomics has aided in deciphering a gene's pattern of DNA and discovering the protein that causes various human disorders. To assess if a disease is decided by a specific gene or numerous genes in the combination of genetic and environmental variables, both traditional genetics and genomics methodologies have been applied (Calvo &Mootha, 2010). We now know why a faulty gene regulates a disease with various degrees of translation (i.e., the image-guided of a gene) or why a gene mutation is generated only in a small group of people (i.e., the expression involved in the gene) but not others thanks to genomics. This answer was developed by discovering the involvement of many other genes in the human genome which link with disease-causing genes to modulate their expression. Sickle cell anemia, Huntington's disease, and cystic fibrosis are all disorders produced by one mutated gene creating one malfunctioning protein (Chambers et al., 2000). Nevertheless, proteomics has revealed why a single gene mutation has many consequences by altering the interacting proteins route, or interactome, which regulates a specific metabolic process. A modification in one metabolic route may potentially have many consequences on the operation of other metabolic processes. Many illnesses, including high blood pressure, diabetes, obesity, high cholesterol, and a variety of cardiovascular and neurological problems, are influenced by a combination of environmental and genetic variables (Bendixen et al., 2010). By defining the functions of diverse proteins and their connections in metabolic processes, the study of proteomics is exposing the involvement of numerous gene bases in illnesses and the impact of environmental variables. Proteomics has also been effective in explaining why some tumors respond to specific drugs while others do not. Proteomics has also helped us know the bad effects of a certain medicine. Proteomics is being utilized to discover specific proteins that might be seen as disease indicators or biomarkers. Proteomics helps determine the metabolic foundation of an illness, as well as diagnose and treat it (Jungblut et al., 1999).

7.4.2 Proteomics of Diagnostic Markers and Drug Development

Proteomics is essential for identifying the etiology and treatment of disease since proteins may serve as both biomarkers and medication targets. Henry Bence-Jones discovered a protein in 1847; this protein was subsequently dubbed the Bence-Jones protein. This finding sparked the concept that proteins may be utilized as biomarkers for human illnesses (He & Chiu, 2003). Kyle found it in 1994 as a loose antibody light chain which is made in overabundance by cancers; due to its lower size, this protein may go past the kidney and is detected in the urine of myeloma patients. That protein was also detected in the serum of individuals with myeloma. As soon as possible, an immunodiagnostic test was created to assess this protein's concentration in cancer patients (Lee et al., 2011). Calculating the volume of Bence-Jones protein has been authorized by the Food and Medication Administration (FDA) and is frequently used to identify myeloma or evaluate the effectiveness of drug therapy in patients. As the patient's situation is stable about a direct therapeutic therapy, this protein diminishes. Numerous proteins have been discovered as potential disease indicators; nevertheless, the majority of those have no specificity and have little clinical utility in clinical practice. Few proteins have been authorized by the FDA as biomarkers for various human illnesses (Hanash, 2003).

The loss of faith in utilizing a specific protein as a biomarker for a disease has resulted in the creation of an array of proteins as biomarkers for specific disorders, rather than a single protein. It has been demonstrated that an elevation of four protein levels, including leptin, prolactin, osteopontin, and insulin-like growth factor II, is a reliable indication of ovarian cancer. That neither of these proteins alone can act as a diagnostic biomarker for ovarian cancer. In combination with proteins, numerous hormones including adrenocorticotropin hormone (ACTH), human chorionic gonadotropin (hCG), and calcitonin are utilized as cancer biomarkers (Hu et al., 2016).

For a variety of disorders, proteins have been utilized as medications or have been the focus of several treatments. Insulin is the prototypical instance of a protein consumed as a medication to regulate blood sugar levels in diabetics. Blood clotting components and immunoglobulins are instances of proteins used as pharmaceuticals. Medications that affect the amount of the protein leptin may be utilized to treat obesity. Some tumors have been successfully treated with Gleevec, a particular inhibitor of kinase implicated in the cell cycle. Similarly, Herceptin (Genentech Inc., South San Francisco, CA) has been used to treat people with cancer (Banks et al., 2000).

Drug discovery is a scientific discipline that employs genomes, proteomics, metabolomics, bioinformatics, and structural chemistry including X-ray crystallography, synthetic chemistry, pharmacology, microbiology, biotechnology, and medical sciences. The first stage in the synthesis of medicine is to identify the source of an illness, which genomics, proteomics, and metabolomics can achieve with relative simplicity. By finding the faulty gene behind a specific illness, genomics may pinpoint the etiology of the sickness. Proteomics can identify the type of faulty protein. Metabolomics may shed light on the biochemical interaction between genes and reveal how the protein products of genes have been altered in a certain illness. Including species devoid of genetic influences, metabolomics may identify the disruption of metabolic pathways induced by environmental stimuli (Jain, 2000). Understanding the structure of proteins and their interactions is crucial for determining the origin of an illness and developing medicine for its treatment or amelioration. Proteins are the reason for an illness, the medications for the sickness, or the target of the pharmaceuticals used to cure the disease; proteomics is essential to comprehending a disease and its therapy (Martin 1999). Bioinformatics is utilized in the synthesis of medications and their ultimate choice as a candidate drug, that is subsequently subjected to biochemical and toxicological testing in model animal systems, followed by human testing, before FDA clearance. Due to the intricacy of the problem and the various lines of approach comprising various scientific disciplines, medication development is a long, multimillion-dollar, multiyear procedure (Aslam et al., 2017).

7.5 METAPROTEOMICS AND HUMAN HEALTH

The DNA pattern of many microbes accessible as culturing in the lab is now available due to breakthroughs in genomics. In gene banks as well as protein databanks, the DNA pattern of these farmed microbes and the translated proteins are combined. Nevertheless, since most microbial organisms cannot be cultured, this figure means just around 1% of entire microbes. As a result, analyzing the mixture existing as a population in a specific habitat is the only approach to obtaining genomic and proteomic data on such bacteria (Mao & Franke, 2015). Metagenomics and meta-proteomics are terms for genomic and proteomic investigations of a population of microorganisms. The earliest investigation of a community of microbes used 16SrRNA analysis, which was followed by DNA sequencing techniques after pyrosequencing technology got accessible. Microorganism metagenomics is progressed, and it has aided in the classification of microbes and the allocation of a protein to specific microbes as quickly as it is detected by proteomics (Zhang et al., 2017). Nevertheless, 2D gel and spectrometric approaches are just now being used in microorganism metaproteomics. Metaproteomics research is more significant than metagenomics research. The research of metagenomics merely reveals the existence of a gene; however, the mere presence of a gene does not guarantee its activation, which determines how a phenotypic manifests. Proteomics, on the other hand, may offer a mapping of metabolic processes in species at a certain moment and in a particular setting. Metaproteomics research is therefore critical for understanding phenotypic manifestation. Furthermore, metaproteomics is critical not just due to many bacterial groups cannot be produced as pure cultures, but because such various organisms collaborate to establish phenotypes. As a result, their gene expression should be investigated in tandem (Long et al., 2020).

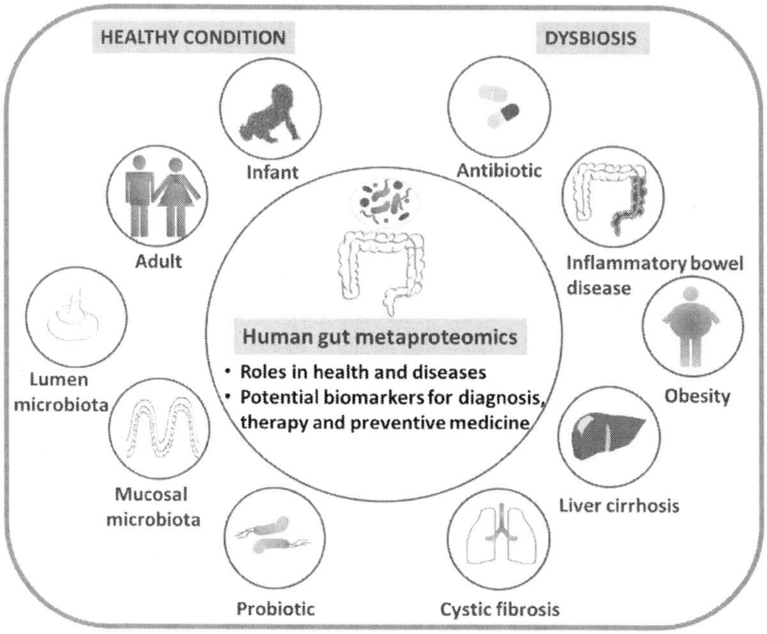

Figure 7.8. Human gut metaproteomic and its applications.

Source: Lee, Pey & Chin, Siok-Fong & Neoh, Hui-min & Jamal, Rahman. (2017). Metaproteomic analysis of human gut microbiota: Where are we heading?. Journal of Biomedical Science. 24. 10.1186/s12929-017-0342-z.

Since many microbes live in symbiotic relationships with humans, metaproteomics research is crucial. Metaproteomics is now the sole approach to comprehending the many metabolic systems and their connections with one another and also with the host's biochemistry, which includes the identification of a specific phenotype (Petriz& Franco, 2017). The human gut microbiome, oral flora, and the urine community of bacteria are instances of this. In humans, the relationship between the microbial community has been thoroughly established. The proportion of distinct microbiome elements changes greatly between normal as well as obese people, indicating that the human gut flora differs greatly. The proportion of Bacteroides to firmicutes typically changes between 0.26 to 1.36; obese people have a lower proportion of Bacteroides to firmicutes in their gut flora (Verberkmoes et al., 2009). Such proportion, though, maybe changed by fat people lowering their calorie consumption. The gut flora also affects insulin resistance in humans, perhaps preventing the formation of type 2 diabetes. Variations in choline metabolism by microorganisms in the human gut appear to have an impact on this. Furthermore, microflora secondary byproducts are beneficial in 2 directions: they provide nutrition to the host and they influence the usage of numerous medications and also their combinations with specific treatments (Zhang et al., 2018). As a result, microflora plays a crucial role in the synthesis of some medications. The elements of microflora vary amongst populations and ethnic communities, suggesting that the host genome has a part in shaping the composition of the host microflora. Such variations in microflora composition have been linked to the results of a proteome investigation of human intestinal flora employing a 2D gel and spectrometry (Zhang &Figeys, 2019).

REFERENCES

1. Aslam, B., Basit, M., Nisar, M. A., Khurshid, M., &Rasool, M. H. (2017). Proteomics: technologies and their applications. *Journal of chromatographic science*, 55(2), 182-196.
2. Banks, R. E., Dunn, M. J., Hochstrasser, D. F., Sanchez, J. C., Blackstock, W., Pappin, D. J., & Selby, P. J. (2000). Proteomics: new perspectives, new biomedical opportunities. *The Lancet*, 356(9243), 1749-1756.
3. Barabási, A. L. (2007). Network medicine—from obesity to the "diseasome". New *England Journal of Medicine*, 357(4), 404-407.
4. Bendixen, E., Danielsen, M., Larsen, K., &Bendixen, C. (2010). Advances in porcine genomics and proteomics—a toolbox for developing the pig as a model organism for molecular biomedical research. *Briefings in functional genomics*, 9(3), 208-219.
5. Calvo, S. E., &Mootha, V. K. (2010). The mitochondrial proteome and human disease. *Annual review of genomics and human genetics*, 11(1), 25-44.
6. Cash, P. (2000). Proteomics in medical microbiology. *Electrophoresis: An International Journal*, 21(6), 1187-1201.
7. Chambers, G., Lawrie, L., Cash, P., & Murray, G. I. (2000). Proteomics: a new approach to the study of disease. *The Journal of pathology*, 192(3), 280-288.
8. Channaveerappa, D., Panama, B. K., &Darie, C. C. (2019). Mass Spectrometry Based Comparative Proteomics Using One Dimensional and Two Dimensional SDS-PAGE of Rat Atria Induced with Obstructive Sleep Apnea. *In Advancements of Mass Spectrometry in Biomedical Research* (Vol. 1, pp. 541-561). Springer, Cham.
9. Cuervo, P., Domont, G. B., & De Jesus, J. B. (2010). Proteomics of trypanosomatids of human medical importance. *Journal of Proteomics*, 73(5), 845-867.
10. Ding, C., Li, Y., Guo, F., Jiang, Y., Ying, W., Li, D., ... & He, F. (2016). A cell-type-resolved liver proteome. *Molecular & Cellular Proteomics*, 15(10), 3190-3202.
11. Goh, K. I., & Choi, I. G. (2012). Exploring the human diseasome: the human disease network. *Briefings in functional genomics*, 11(6), 533-542.

12. Hanash, S. (2003). Disease proteomics. *Nature*, 422(6928), 226-232.

13. He, F. (2005). Human liver proteome project. *Molecular & Cellular Proteomics*, 4(12), 1841-1848.

14. He, Q. Y., & Chiu, J. F. (2003). Proteomics in biomarker discovery and drug development. *Journal of cellular biochemistry*, 89(5), 868-886.

15. Hu, S., Loo, J. A., & Wong, D. T. (2006). Human body fluid proteome analysis. *Proteomics*, 6(23), 6326-6353.

16. Hu, S., Yen, Y., Ann, D., & Wong, D. T. (2007). Implications of salivary proteomics in drug discovery and development: a focus on cancer drug discovery. *Drug discovery today*, 12(21-22), 911-916.

17. Huang, L., Shao, D., Wang, Y., Cui, X., Li, Y., Chen, Q., & Cui, J. (2021). Human body-fluid proteome: quantitative profiling and computational prediction. *Briefings in bioinformatics*, 22(1), 315-333.

18. HW Dekkers, D., Bezstarosti, K., Kuster, D., JM Verhoeven, A., & K Das, D. (2010). Application of proteomics in cardiovascular research. *Current Proteomics*, 7(2), 108-115.

19. Jain, K. K. (2000). Applications of proteomics in oncology. *Pharmacogenomics*, 1(4), 385-393.

20. Jungblut, P. R., Zimny-Arndt, U., Zeindl-Eberhart, E., Stulik, J., Koupilova, K., Pleißner, K. P., ... &Stöffler, G. (1999). Proteomics in human disease: cancer, heart and infectious diseases. *Electrophoresis: An International Journal*, 20(10), 2100-2110.

21. Klose, J., Nock, C., Herrmann, M., Stühler, K., Marcus, K., Blüggel, M., ... &Lehrach, H. (2002). Genetic analysis of the mouse brain proteome. *Nature genetics*, 30(4), 385-393.

22. Kuscuoglu, D., Janciauskiene, S., Hamesch, K., Haybaeck, J., Trautwein, C., &Strnad, P. (2018). Liver–master and servant of serum proteome. *Journal of Hepatology*, 69(2), 512-524.

23. Lee, J. M., Han, J. J., Altwerger, G., & Kohn, E. C. (2011). Proteomics and biomarkers in clinical trials for drug development. *Journal of proteomics*, 74(12), 2632-2641.

24. Liotta, L. A., Ferrari, M., &Petricoin, E. (2003). Clinical proteomics: written in blood. *Nature*, 425(6961), 905-905.

25. Long, S., Yang, Y., Shen, C., Wang, Y., Deng, A., Qin, Q., &Qiao, L. (2020). Metaproteomics characterizes human gut microbiome function in colorectal cancer. *NPJ biofilms and microbiomes,* 6(1), 1-10.

26. Macaulay, I. C., Carr, P., Gusnanto, A., Ouwehand, W. H., Fitzgerald, D., & Watkins, N. A. (2005). Platelet genomics and proteomics in human health and disease. *The Journal of clinical investigation*, 115(12), 3370-3377.

27. Mao, L., & Franke, J. (2015). Symbiosis, dysbiosis, and rebiosis—The value of metaproteomics in human microbiome monitoring. *Proteomics*, 15(5-6), 1142-1151.

28. Midic, U., Oldfield, C. J., Dunker, A. K., Obradovic, Z., &Uversky, V. N. (2009). Protein disorder in the human diseasome: unfoldomics of human genetic diseases. *BMC genomics*, 10(1), 1-24.

29. Mouradian, S. (2002). Lab-on-a-chip: applications in proteomics. *Current opinion in chemical biology*, 6(1), 51-56.

30. Pedret, A., Catalán, U., Rubió, L., Baiges, I., Herrero, P., Piñol, C., ... & Solà, R. (2021). Phosphoproteomic Analysis and Protein–Protein Interaction of Rat Aorta GJA1 and Rat Heart FKBP1A after Secoiridoid Consumption from Virgin Olive Oil: A Functional Proteomic Approach. *Journal of Agricultural and Food Chemistry*, 69(5), 1536-1554.

31. Petricoin, E. F., &Liotta, L. A. (2003). Clinical applications of proteomics. *The Journal of nutrition*, 133(7), 2476S-2484S.

32. Petriz, B. A., & Franco, O. L. (2017). Metaproteomics as a complementary approach to gut microbiota in health and disease. *Frontiers in chemistry*, Vol. 5, pp. 1-4.

33. Sharma, K., Schmitt, S., Bergner, C. G., Tyanova, S., Kannaiyan, N., Manrique-Hoyos, N., ... & Simons, M. (2015). Cell type–and brain region–resolved mouse brain proteome. *Nature neuroscience*, 18(12), 1819-1831.

34. Shi, R., Kumar, C., Zougman, A., Zhang, Y., Podtelejnikov, A., Cox, J., ... & Mann, M. (2007). Analysis of the mouse liver proteome using advanced mass spectrometry. *Journal of proteome research*, 6(8), 2963-2972.

35. Sung, H. J., Ryang, Y. S., Jang, S. W., Lee, C. W., Han, K. H., &Ko, J. (2006). Proteomic analysis of differential protein expression in atherosclerosis. *Biomarkers*, 11(3), 279-290.

36. Verberkmoes, N. C., Russell, A. L., Shah, M., Godzik, A., Rosenquist, M., Halfvarson, J., ... &Jansson, J. K. (2009). Shotgun metaproteomics of the human distal gut microbiota. *The ISME journal*, 3(2), 179-189.

37. Wysocki, K., & Ritter, L. (2011). Diseasome. *Annual review of nursing research*, 29(1), 55-72.

38. Yokota, H. (2019). Applications of proteomics in pharmaceutical research and development. *Biochimica et Biophysica Acta (BBA)-Proteins and Proteomics*, 1867(1), 17-21.

39. Zhang, X., &Figeys, D. (2019). Perspective and guidelines for metaproteomics in microbiome studies. *Journal of proteome research*, 18(6), 2370-2380.

40. Zhang, X., Chen, W., Ning, Z., Mayne, J., Mack, D., Stintzi, A., ... &Figeys, D. (2017). Deep metaproteomics approach for the study of human microbiomes. *Analytical chemistry*, 89(17), 9407-9415.

41. Zhang, X., Li, L., Mayne, J., Ning, Z., Stintzi, A., &Figeys, D. (2018). Assessing the impact of protein extraction methods for human gut metaproteomics. *Journal of proteomics*, 180(1), 120-127.

CHAPTER 8

DEVELOPMENTS IN PROTEOMICS

CONTENTS

8.1 INTRODUCTION

Proteomics began with the launch of protein cleaning processes. Originally, it consisted primarily of purifying a single protein at a moment by column chromatography. Electrofocusing of proteins was a significant advancement in protein purification that resulted in the creation of a two-dimensional (2D) gel. Numerous proteins were purified on a single two-dimensional gel for further characterization. The advent of mass spectrometry, genomics, and bioinformatics ushered in a new era of proteomics. Today, many proteins may be examined concurrently (Faber et al., 2006).

8.2 TECHNICAL SCOPE OF PROTEOMICS

Protein extraction by 2D gel and/or high-performance column chromatography (HPLC), and protein recognition by mass spectrometry (MS) in combination with protein data banks as well as a variety of software applications, are currently adequate proteomic approaches. To be more effective, these approaches must be enhanced in the near (Molloy et al., 2003).

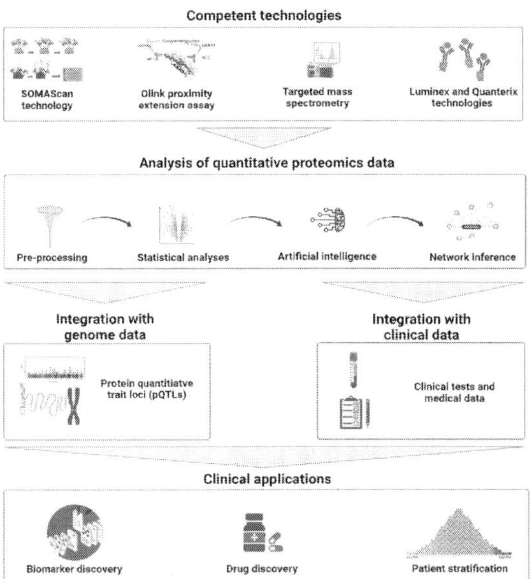

Figure 8.1. Types-of-proteomics-and-their-applications-to-biology.

Source: Correa Rojo A, Heylen D, Aerts J, Thas O, Hooyberghs J, Ertaylan G and Valkenborg D (2021) Towards Building a Quantitative Proteomics Toolbox in Precision Medicine: A Mini-Review. Front. Physiol. 12:723510. doi: 10.3389/fphys.2021.723510

As noted by others, many predicted changes in the coming years must have the coming up strategies (Liska& Shevchenko, 2003):

- To enhance a low-abundance protein by preferentially removing high-abundance proteins. Today, immunoprecipitation is used to concentrate low-abundance proteins due to the elimination of numerous high-abundance proteins. Several growth hormones and peptides, including cytokines, are also eliminated throughout this procedure due to the nonspecific attachment of such substances to abundant proteins. Low- and high-abundance proteins appear to occur with a 10-order magnitude difference. Modern protein extraction methods distinguish between high and low abundance proteins in the individual plasma proteome by three to four orders of magnitude. Numerous low-abundance molecules that are not yet found were amongst such strategies. This kind of signaling pathway and peptides could be prospective biomarkers; hence, ways should be conducted in the future to acquire or isolate them without removing them with the blood's abundant proteins (Cheung et al., 2021).
- The lowering of the platform's size.

The small size of the platform is essential to start reducing the protein analysis procedure, which is extremely detailed with a much-limited sample volume. Eventually, it could be feasible to lower the sample group from a nanoliter to a picoliter, hence decreasing the quantity of protein from females to automobiles. Miniaturization will need the use of microfluidics to manage minuscule samples with no absorption or evaporating waste. Thus, the protein specimen would be able to be analyzed on a protein chip (Dakna et al., 2009).

- Automation.

Utilizing robots to perform programmed isoelectrofocusing (IEF) and 2D gel unit analysis as well as post-sample processing following the separating proteins, progress has been made to automate the protein analytical method. Automation will also include the removal of protein patches from the gel, their breakdown using trypsin, and the subsequent cleaning and placement of a limited sample on the MALDI-TOF-MS port (Figeys& Pinto, 2001).

- New software applications and quicker computers.

It is required to develop new software and faster computers to allow more accurate detection, quantification, and identification of proteins in the gel spot.

Methods will be developed to increase the sensitivity of current instruments to avoid false-positive or false-negative results among MS data. Currently, up to 20% of MS data included such false positives/negatives (Phillips &Bogyo, 2005).

- Heightened sensibility.

As discussed in Chapter 4, new software is needed to improve the accuracy of separating proteins, measurement, and classification in gel spots (Procko, 2020).

- Artificial intelligence (AI).

It might be conceivable to create and use AI to discover and describe proteins. Recent advances in artificial intelligence have made it feasible to find several indicators for ovarian cancer (Demkow, 2010).

- The evolution of in-situ mass spectrometry.

Techniques are now being developed to examine the proteins in tissue slices. To use this method, it is feasible to contrast the protein composition of a healthy cell to that of a cell from such a diseased individual. This one will aid in identifying the disease's protein biomarker(s) for diagnostics, therapeutic discovery, or monitoring the efficacy of therapy throughout treatment (Molloy &Witzmann, 2002).

- Analysis of biological pathways via computer.

Most illnesses include metabolic path connections. Novel computational systems will be able to discover the consequences of metabolic interactions, which will aid in the creation of diagnostic and therapeutic agents (Russell & Lilley, 2012).

8.3 SCIENTIFIC SCOPE OF PROTEOMICS

Epigenomics encompasses both protein and small-molecule modifications to the genome. Knowing epigenomics is the best way to solve crucial biological problems. This is especially true in terms of diversity and growth. This could describe how the style and substance of countless cell types in multicellular organisms with a similar genome vary. Furthermore, cancer as well as other human illnesses, and aging, are caused by epigenetic alterations. Due to their significance in the nucleosomal arrangement of chromosomes, proteins play a unique role in addressing such concerns (Lindsey et al., 2015).

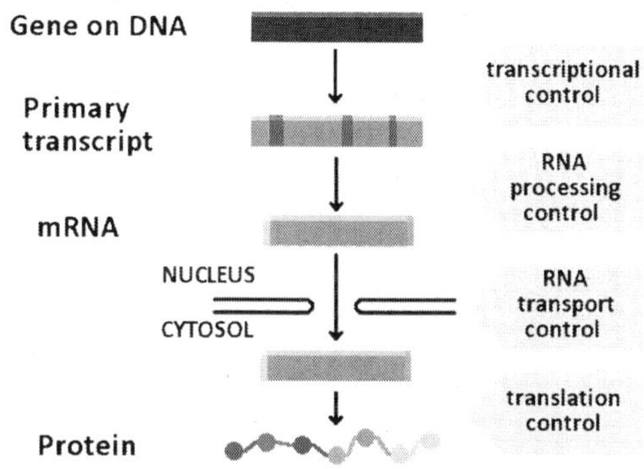

Figure 8.2. Genes, proteins, and molecular machinery.

Source: By ArneLH - Own work, CC BY-SA 3.0, https://commons.wikimedia.org/w/index.php?curid=5815139

Many genes are transcribed whereas others are masked by the nucleosomal arrangement of chromosomes, resulting in the diversification of types of cells and the formation of tissues. Moreover, transcriptionally proteins give an additional aspect to the differentiating of cell types. Shinya Yamanaka and his Japanese colleagues (Takahashi et al., 2007) have transformed ordinary skin cells into pluripotent stem cells by introducing four genes for transcription factors. Shinya Yamanaka and his colleagues then transformed such stem cells to produce an assortment of tissues. These findings presented an option for acquiring human embryonic stem cells, that makes a significant contribution to the controversy involving the utilization of embryonic stem cells by some U.S. organizations. It is evident from this research that several types of transcription factors occur. Such a team of Japanese authors brought transcription parameters into skin cells that correspond to the category of master transcription parameters (Liska& Shevchenko, 2003). This or other research demonstrates conclusively the relevance of proteomics in explaining epigenesis. Epigenesis is gaining importance from both a scientific and an applied standpoint: Based on a knowledge of the epigenetic modifications of genes, several diagnostics have been created. In the near, it is anticipated as proteomics may give hints for the diversification of cell types from stem cells and their usage in various forms of treatments for various human illnesses. Thus, it is anticipated that proteomics will soon give a comprehensive knowledge of the difficulties of diversification throughout human evolution, as well as its involvement in the generation and utilization of stem cells (Tsiamis et al., 2019).

8.4 MEDICAL SCOPE OF PROTEOMICS

The proteomic industry has brought researchers a better understanding of the expression of proteins, such as their many alterations, connections, and functions in metabolic processes. It is intended to give light on the reasons for many illnesses, as well as how to diagnose, treat, and cure them. This chapter goes through some of these components (Xu et al., 2020).

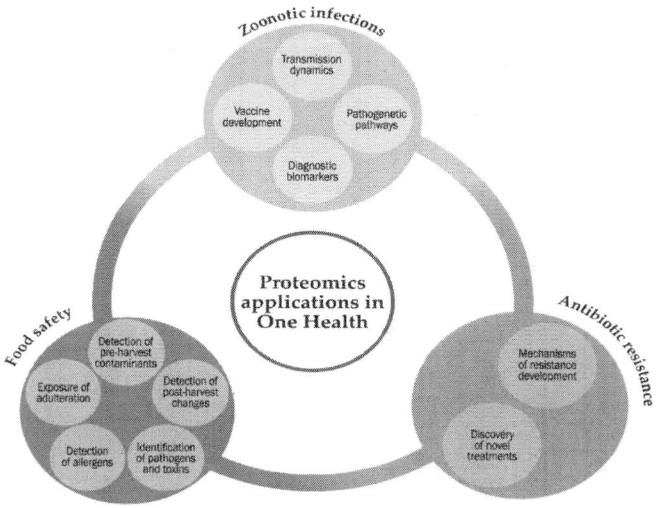

Figure 8.3. An overview of how proteomics methods are used in One Health.

Source: Katsarou EI, Billinis C, Galamatis D, Fthenakis GC, Tsangaris GT, Katsafadou AI. Applied Proteomics in 'One Health'. Proteomes. 2021; 9(3):31. https://doi.org/10.3390/proteomes9030031

8.4.1 Human Diseases

Proteomics has significant potential for the prevention and therapy of cancer, cardiovascular, neurodegenerative, metabolic, and viral illnesses, among others. Proteomics has enabled the development of a panel of biomarkers for cancer diagnosis. The majority of such screens contain numerous proteins, often exceeding 25 proteins. In the years ahead, this might be feasible to select fewer proteins as important biomarkers. Utilizing AI technology and enhanced software tools, it would be feasible to identify the important indicators (Risch &Merikangas, 1996). It'd be feasible to diagnose cancer at an earlier, benign phase, making therapy far more bearable. It might be conceivable to discover biomarkers that differentiate between aggressive and slowly developing cancer. Amongst some of the proteins for sending and receiving messages, protein complexes, regulators of metabolic paths, and switches, a few such indicators would've been discovered. Similarly, markers for the early identification of cardiovascular, neurological, and metabolic illnesses will be developed and utilized to diagnose and treat illness. Proteomic research is predicted to improve the early diagnosis of infectious illnesses, much as parasite proteomics is improving the identification of malaria (Matés et al., 1999).

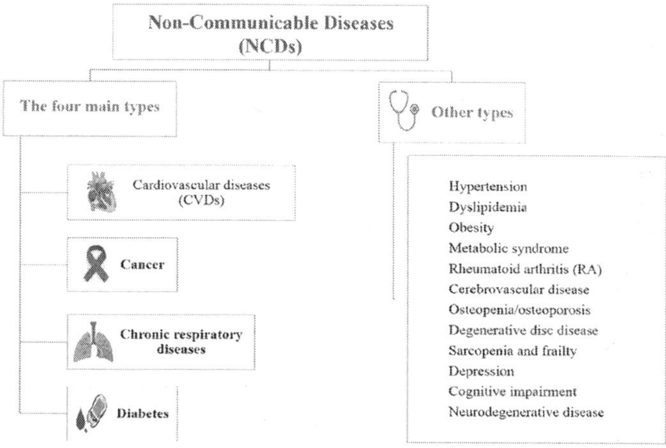

Figure 8.4. Schematic displaying the prevalent NCDs in the body.

Source: Budreviciute A, Damiati S, Sabir DK, Onder K, Schuller-Goetzburg P, Plakys G, Katileviciute A, Khoja S and Kodzius R (2020) Management and Prevention Strategies for Non-communicable Diseases (NCDs) and Their Risk Factors. Front. Public Health 8:574111. doi: 10.3389/fpubh.2020.574111

Studying the proteome of the host following an assault by a bacteria should offer insight into the treatment of antibiotic-resistant bacteria (Malech &Gallin, 1987). Knowing the genomes and proteomics of viruses might give insight into how to cure or avoid viral infections like severe acute respiratory syndrome (SARS) and the common cold. The genomes of over 97 cold viruses have enabled the identification of a common area in the whole of them, which would be attacked by a medicine in the near to give more effective treatment for the cold virus (Wilson et al., 2007).

8.4.2 Development of Drugs

The capacity to create better medications for various infections is the goal of proteomics. Proteomics is thought to be a safer test to treat, cure, and analyze the performance of pharmacological therapy. It is envisaged that biomarkers discovered via proteomic studies of healthy and sick individuals would be used to develop improved diagnoses and smarter medications in the area of proteomics. Smart medications will be developed by comparing the proteomes of healthy cells and cells from a sick individual and identifying protein biomarkers, their alteration, and changed metabolic pathways (Sena et al., 2007). A new drug is a costly endeavor. Proteomics is predicted to lower costs by expanding the range of protein targets employed in medication development. With the use of modern bioinformatics tools, understanding metabolic processes and protein connections would be exploited to aid in the cost-effective production of medications. The combination of potential compounds as medications with protein targets would be accessible quickly and cost-effectively in the future using high throughput screening (HTS) technologies. The utilization of combinatorial chemistry and the chemical collection accessible on the web will help this method (Steverding, 2010).

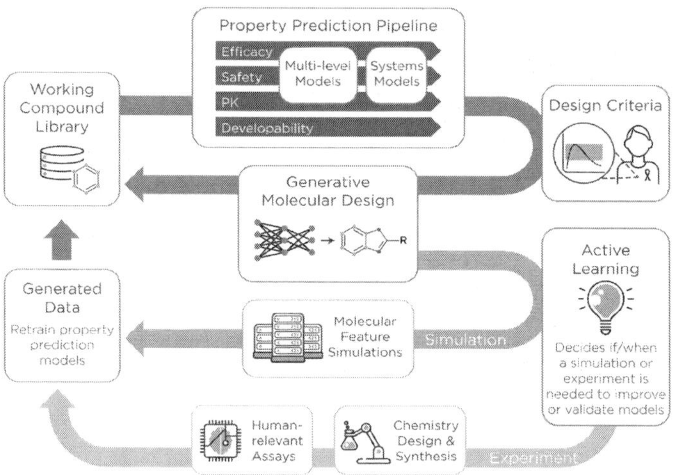

Figure 8.5. Drug Discovery Steps.

Source: Hinkson IV, Madej B and Stahlberg EA (2020) Accelerating Therapeutics for Opportunities in Medicine: A Paradigm Shift in Drug Discovery. Front. Pharmacol. 11:770. doi: 10.3389/fphar.2020.00770

Proteomics would also assist in the identification of novel proteins that can be used as pharmaceuticals. The research of Mark Tuszynski and his colleagues at the University of California, San Diego, suggests this idea (Winters &Arria, 2011). The researchers discovered that injecting a brain-derived neurotrophic factor (BDNF) into mice, rhesus monkeys, and other model animals may alleviate signs of Alzheimer's disease-like memory loss, brain cell degeneration, and cognitive impairment. The BDNF protein is normally synthesized in the brain of healthy animals, although it is absent from the cortex of diseased animals. Proteomics is predicted to aid in the creation of vaccinations in the years ahead, other than a medicine production (Kim et al., 2013).

8.4.3 Personalized Medicine

The goal is that proteomics will create customized medicine a standard component of clinical practice as well as care delivery. With the advancements in genomes and proteomics, it is conceivable to build diagnostic algorithms that predict the optimal person for a medicine prescription. It appears that a minimum of 200 recommended medications are ineffective or have negative consequences for a given patient (Jain, 2002). In this kind of instance, the patient possesses a gene mutation that renders a specific medicine ineffective. The antiretroviral medication abacavir, for instance, is ineffective in 20% of individuals who possess a mutation in the HLA gene. That a screening test is sufficient to assess the HLA variation in a population, abacavir may be prescribed to just those individuals who do not have HLA mutation and would advantage from its usage. Similarly, not all individuals may be benefited from warfarin, an anticoagulant utilized to help patients with thrombosis and other blood diseases (Mura &Couvreur, 2012). To prevent internal bleeding in this kind of individual, the warfarin dosage should be modified. A diagnostic test for warfarin is developed, but the U.S. Food and Drug Administration has not yet authorized it (FDA). It is recognized that some cancer patients never recover from tamoxifen since they contain a mutation in the 2D6 gene which generates enzymes that transforms tamoxifen into endoxifen, a chemical having cancer-fighting properties. Such breast cancer patients must take aromatase inhibitors, a different family of medications. However, aromatase inhibitors have a restricted use since they are only effective in postmenopausal women. This provides the lady with breast cancer who still hasn't reached menopause with a variety of unusual treatment options.

Such patients need the development of new medications (Chan & Ginsburg, 2011). Similarly, it has been said that some colon cancer patients with the KRAS mutation never recover from the prescription of Vectibix (Amgen, Inc., Thousand Oaks, CA). Currently, Vectibix is exclusively administered to individuals without the KRAS mutation. In the years ahead, it will be commonplace to run personalized genomics on a person or to need a specific diagnostic test before prescribing a medicine (Ginsburg & Willard, 2009). The administration of tailored medication will need the construction of an electronic repository for an individual's whole health information and drug reactions depending on many diagnostics, subgroups of illness, and genetics. Related to the creation of diagnostic tests for using pharmaceuticals and the expenditure on producing these diagnostics, the technique of personalized medicine presents several medical ethical concerns. It also includes issues about the prescription of specific treatment regimens by doctors (Joyner &Paneth, 2015).

8.4.4 Proteomics and Metabolomics

The research of metabolites is known as metabolic. Metabolites are chemicals produced by the enzymatic activity of proteins along any metabolic route (Titz et al., 2014). It is recognized as genes make enzymes so these enzymes accelerate a biochemical to form metabolites, that define an organism's phenotype. Consequently, genomes, proteomics, and metabolomics are interconnected. Indicators of illness conditions and metabolites contribute to the development of diagnoses for common diseases (Horgan& Kenny, 2011). The existence of sarcosine in urine samples has been identified as a sign of an aggressive type of prostate cancer. One type of human prostate cancer is often slow-growing, whilst the other is typically fast-growing. The species with rapid growth is invasive. Therefore, the identification of sarcosine in urine samples may be utilized to diagnose the aggressive type of prostate cancer at a preliminary phase, allowing for appropriate treatment (Yizhak et al., 2010). The detection of sarcosine in urine may be used to determine the proteome alterations in people with cancer. The existence of sarcosine implies a deficiency in the enzyme which transforms glycine to sarcosine or alterations in the protein(s) which catalyze sarcosine breakdown. Research from the University of California, Los Angeles demonstrates further impacts of molecules and metabolites on proteomics (UCLA) (Kovac et al., 2013).

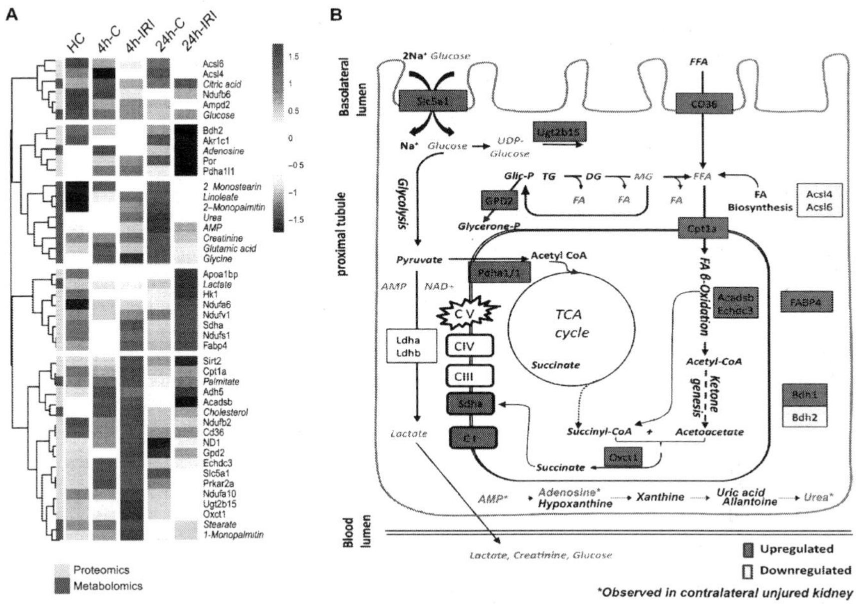

Figure 8.6. Integrated-proteomic-and-metabolomic-analysis 4-h and 24-h post-IRI-A Metabolites.

Source: Huang, Hong lei & Dullemen, Leon & Akhtar, M Zeeshan & Lo Faro, Maria Letizia & Yu, Zhanru & Valli, Alessandro & Dona, Anthony & Thézénas, Marie-Laëtitia & Charles, Philip & Fischer, Roman & Kaisar, Maria & Leuvenink, Henri & Ploeg, Rutger & Kessler, Benedikt. (2018). Proteo-metabolomics reveals compensation between ischemic and non-injured contralateral kidneys after reperfusion. Scientific Reports. 8. 10.1038/s41598-018-26804-8.

The UCLA researchers have demonstrated that green tea affects the proteins associated with cell movement by altering the proteome of lung cells. These researchers conclude that polyphenols in green tea extract alter the proteome of lung tissue and are accountable for suppressing cancer cell spread and proliferation (Lindenburg et al., 2015). Similarly, it has been demonstrated that teenage alcohol intake produces proteome alterations in the hippocampal area of the brain. Such research gives the biochemical foundation for complementary and alternative medicine, as well as the connections between metabolomics, proteomics, and human health (Lange et al., 2008).

The metabolomics method for disease detection is straightforward, noninvasive, and far less costly than genomic and proteomic approaches. By analyzing the metabolic processes both in healthy and diseased cells, Israeli researchers have built computer systems that monitor multiple metabolites. A repository of metabolites has become accessible as a database. The study of metabolomics would identify a significant portion of proteomics (Zhao & Lin, 2014).

8.5 PROTEOMICS, ENERGY PRODUCTION, AND BIOREMEDIATION

All types of energy on Earth are fundamentally bioenergy, and that is solar energy captured via biological processes. Consequently, biology, like engineering and chemistry, is essential for solving our energy concerns. Presently, research is solely engaged in the generation of biofuels, the filtration of water, as well as other microbial community-based environmental issues (Zhao &Poh, 2008). Future developments in synthetic genomes and proteomics are anticipated to address our power generation, bioremediation of the ecosystem, carbon sequestration, as well as other issues related. The goal of synthetic genomics is to generate new microorganisms with carefully designed genomes that are suited for energy generation and ecological bioremediation. The age of synthetic proteomics will shortly be inaugurated by the proteomics of such new fully species (Wilkins et al., 2009).

8.6 PROTEOMICS AND BIODEFENSE

The rising problem of bioterrorism and the ongoing development of new contagious diseases have prompted a significant resurgence in scientific research attempts to expand better treatments, diagnostics, and vaccines, and to boost the basic knowledge of the host's immune system to bacterial infections. The accessibility of numerous mass spectrometry systems, in conjunction with multivariate extraction techniques and microbiological genetic databases, offers a unique opportunity to build these vital resources (Drake et al., 2005). Presents a summary of existing proteomic methodologies used on bacteria and viruses regarded as potential causes of bioterrorism. In addition, the use of immunoproteomics in the creation of novel vaccine targets has been reviewed. These effective research methods may provide a huge number of possible novel protein targets; nevertheless, the translation of these proteomic findings into practical anti-bioterrorism therapies will need extensive collaborative research across several fundamental scientific and clinical fields. Utilizing the influenza virus as an instance, a paradigm for translational proteomic research is presented to illustrate this method (Demirev &Fenselau, 2008).

Now since the world trade center attacks on September 11, 2001, the concept of biodefense became an actuality. It is anticipated that numerous diagnostics and vaccinations against a variety of microorganisms will be produced due in great part to the Departments of Defense and Homeland Security's grasp of proteomics (Whitesides &Stroock, 2001).

REFERENCES

1. Chan, I. S., & Ginsburg, G. S. (2011). Personalized medicine: progress and promise. *Annual review of genomics and human genetics*, 12, 217-244.
2. Cheung, T. K., Lee, C. Y., Bayer, F. P., McCoy, A., Kuster, B., & Rose, C. M. (2021). Defining the carrier proteome limit for single-cell proteomics. *Nature methods*, 18(1), 76-83.
3. Dakna, M., He, Z., Yu, W. C., Mischak, H., &Kolch, W. (2009). Technical, bioinformatical and statistical aspects of liquid chromatography–mass spectrometry (LC–MS) and capillary electrophoresis-mass spectrometry (CE-MS) based clinical proteomics: A critical assessment. *Journal of Chromatography B*, 877(13), 1250-1258.
4. Demirev, P. A., &Fenselau, C. (2008). Mass spectrometry in biodefense. *Journal of mass spectrometry*, 43(11), 1441-1457.
5. Demkow, U. (2010). Laboratory medicine in the scope of proteomics and genomics. *EJIFCC*, 21(3), pp. 56.
6. Drake, R. R., Deng, Y., Schwegler, E. E., &Gravenstein, S. (2005). Proteomics for biodefense applications: progress and opportunities. *Expert Review of Proteomics*, 2(2), 203-213.
7. Faber, M. J., Agnetti, G., Bezstarosti, K., Lankhuizen, I. M., Dalinghaus, M., Guarnieri, C., ... &Lamers, J. M. (2006). Recent developments in proteomics. *Cell biochemistry and biophysics*, 44(1), 11-29.
8. Figeys, D., & Pinto, D. (2001). Proteomics on a chip: promising developments. *Electrophoresis*, 22(2), 208-216.
9. Ginsburg, G. S., & Willard, H. F. (2009). Genomic and personalized medicine: foundations and applications. *Translational research*, 154(6), 277-287.
10. Horgan, R. P., & Kenny, L. C. (2011). 'Omic'technologies: genomics, transcriptomics, proteomics and metabolomics. *The Obstetrician &Gynaecologist*, 13(3), 189-195.
11. Jain, K. K. (2002). Personalized medicine. *Current opinion in molecular therapeutics*, 4(6), 548-558.
12. Joyner, M. J., &Paneth, N. (2015). Seven questions for personalized medicine. *JAMA*, 314(10), 999-1000.
13. Kim, J. H., O'Brien, K. M., Sharma, R., Boshoff, H. I., Rehren, G., Chakraborty, S., ... &Schnappinger, D. (2013). A genetic strategy to

identify targets for the development of drugs that prevent bacterial persistence. *Proceedings of the National Academy of Sciences*, 110(47), 19095-19100.

14. Kovac, J. R., Pastuszak, A. W., & Lamb, D. J. (2013). The use of genomics, proteomics, and metabolomics in identifying biomarkers of male infertility. *Fertility and sterility*, 99(4), 998-1007.
15. Lange, E., Tautenhahn, R., Neumann, S., &Gröpl, C. (2008). Critical assessment of alignment procedures for LC-MS proteomics and metabolomics measurements. *BMC Bioinformatics*, 9(1), 1-19.
16. Lindenburg, P. W., Haselberg, R., Rozing, G., &Ramautar, R. (2015). Developments in interfacing designs for CE–MS: towards enabling tools for proteomics and metabolomics. *Chromatographia*, 78(5), 367-377.
17. Lindsey, M. L., Mayr, M., Gomes, A. V., Delles, C., Arrell, D. K., Murphy, A. M., ... & Srinivas, P. R. (2015). Transformative impact of proteomics on cardiovascular health and disease: a scientific statement from the American Heart Association. *Circulation*, 132(9), 852-872.
18. Liska, A. J., & Shevchenko, A. (2003). Expanding the organismal scope of proteomics: cross-species protein identification by mass spectrometry and its implications. *Proteomics*, 3(1), 19-28.
19. Malech, H. L., &Gallin, J. I. (1987). Neutrophils in human diseases. *New England Journal of Medicine*, 317(11), 687-694.
20. Matés, J. M., Pérez-Gómez, C., & De Castro, I. N. (1999). Antioxidant enzymes and human diseases. *Clinical biochemistry*, 32(8), 595-603.
21. Molloy, M. P., &Witzmann, F. A. (2002). Proteomics: technologies and applications. *Briefings in Functional Genomics*, 1(1), 23-39.
22. Molloy, M. P., Brzezinski, E. E., Hang, J., McDowell, M. T., &VanBogelen, R. A. (2003). Overcoming technical variation and biological variation in quantitative proteomics. *Proteomics*, 3(10), 1912-1919.
23. Mura, S., &Couvreur, P. (2012). Nanotheranostics for personalized medicine. *Advanced drug delivery reviews,* 64(13), 1394-1416.
24. Phillips, C. I., &Bogyo, M. (2005). Proteomics meets microbiology: technical advances in the global mapping of protein expression and function. *Cellular microbiology*, 7(8), 1061-1076.

25. Procko, E. (2020). Deep mutagenesis in the study of COVID-19: a technical overview for the proteomics community. *Expert review of proteomics*, 17(9), 633-638.

26. Risch, N., &Merikangas, K. (1996). The future of genetic studies of complex human diseases. *Science*, 273(5281), 1516-1517.

27. Russell, M. R., & Lilley, K. S. (2012). Pipeline to assess the greatest source of technical variance in quantitative proteomics using metabolic labelling. *Journal of proteomics*, 77, 441-454.

28. Sena, E., van der Worp, H. B., Howells, D., & Macleod, M. (2007). How can we improve the pre-clinical development of drugs for stroke?. *Trends in neurosciences*, 30(9), 433-439.

29. Steverding, D. (2010). The development of drugs for treatment of sleeping sickness: a historical review. *Parasites & vectors*, 3(1), 1-9.

30. Takahashi, D., Li, B., Nakayama, T., Kawamura, Y., &Uemura, M. (2013). Plant plasma membrane proteomics for improving cold tolerance. *Frontiers in plant science*, 4(1), pp. 90.

31. Titz, B., Elamin, A., Martin, F., Schneider, T., Dijon, S., Ivanov, N. V., ... &Peitsch, M. C. (2014). Proteomics for systems toxicology. *Computational and Structural Biotechnology Journal*, 11(18), 73-90.

32. Tsiamis, V., Ienasescu, H. I., Gabrielaitis, D., Palmblad, M., Schwämmle, V., &Ison, J. (2019). One thousand and one software for proteomics: tales of the toolmakers of science. *Journal of proteome research*, 18(10), 3580-3585.

33. Whitesides, G. M., &Stroock, A. D. (2001). Flexible methods for microfluidics. *Phys. Today*, 54(6), 42-48.

34. Wilkins, M. J., VerBerkmoes, N. C., Williams, K. H., Callister, S. J., Mouser, P. J., Elifantz, H., ... &Banfield, J. F. (2009). Proteogenomic monitoring of Geobacter physiology during stimulated uranium bioremediation. *Applied and environmental microbiology*, 75(20), 6591-6599.

35. Wilson, A. S., Power, B. E., & Molloy, P. L. (2007). DNA hypomethylation and human diseases. *Biochimica et Biophysica Acta (BBA)-Reviews on Cancer*, 1775(1), 138-162.

36. Winters, K. C., &Arria, A. (2011). Adolescent brain development and drugs. *The prevention researcher*, 18(2), pp. 21.

37. Xu, Y., Li, X., Man, D., & Su, X. (2020). iTRAQ-based proteomics analysis on insomnia rats treated with Mongolian medical warm acupuncture. *Bioscience reports*, 40(5), pp. 1-6.
38. Yizhak, K., Benyamini, T., Liebermeister, W., Ruppin, E., &Shlomi, T. (2010). Integrating quantitative proteomics and metabolomics with a genome-scale metabolic network model. *Bioinformatics*, 26(12), i255-i260.
39. Zhao, B., &Poh, C. L. (2008). Insights into environmental bioremediation by microorganisms through functional genomics and proteomics. *Proteomics*, 8(4), 874-881.
40. Zhao, Y. Y., & Lin, R. C. (2014). UPLC–MSE application in disease biomarker discovery: the discoveries in proteomics to metabolomics. *Chemico-biological interactions*, 215, 7-16.

INDEX